REAL ANALYSIS

REAL ANALYSIS

S. Nanda
V.P. Saxena

ALLIED PUBLISHERS LIMITED

New Delhi • Mumbai • Calcutta • Lucknow • Chennai
Nagpur • Bangalore • Hyderabad • Ahmedabad

ALLIED PUBLISHERS LIMITED

Regd. Off.: 15. J.N. Heredia Marg, Ballard Estate, Mumbai — 400038
Prarthna Flats (2nd Floor), Navrangpura, Ahmedabad — 380009
3-2-844/6 & 7 Kachiguda Station Road, Hyderabad — 500027
16A Ashok Marg, Patiala House, Lucknow — 226001
5th Main Road, Gandhinagar, Bangalore — 560009
17 Chittaranjan Avenue, Calcutta — 700072
81 Hill Road, Ramnagar, Nagpur — 440010
13/14, Asaf Ali Road, New Delhi — 110002
751 Anna Salai, Chennai — 60000

ISBN 81-7764-062-3
ISBN 13: 978-81-7764-062-5

ASSOCIATE EDITOR	:	JYOTI MEHROTRA
PRODUCTION	:	B.K. DAS
COPY EDITOR	:	D.S. SASTRY
COVER DESIGN	:	KAM STUDIO

Glossary

Absolute Convergence of Series: A series Σa_n is absolutely convergent if $\Sigma |a_n|$ is convergent.

Alternating Series: A series consisting of positive and negative terms.

Archimedean: An ordered field F is Archimedean if for b, c in F, $c > 0$, there exists a positive integer n such that $nc > b$.

Axiom of Completeness: A nonempty set of real numbers which is bounded above has a least upper bound.

Axiom of Order: The field of all real numbers is an ordered field.

Antisymmetric: A relation R is antisymmetric if xRy and yRx $\Rightarrow x = y$.

Bolzano-Weierstrass Theorem: Every infinite bounded set has a limit point.

Bounded Above (Below): A subset of real line is bounded above (below) if it has an upper bound (a lower bound).

Bounded Function: A function $f : R \to R$ is bounded if there is a constant $M > 0$ such that $|f(x)| \leq M$ for all x in the domain.

Bounded Sequence: A sequence $\{x_n\}$ of real numbers is bounded if there is a constant M > 0 such that $|x_n| \leq M$ for all $n = 1, 2, 3, \ldots$.

Bounded Variation: A function f is of bounded variation on $[0, 1]$ if

$$v(f) = \sup \sum_{i=0}^{n} |f(x_{i+1}) - f(x_i)|$$

is finite, where sup is taken over all partitions $0 = x_0 < x_1 < \ldots < x_n = 1$.

Countable Set: A set is countable if it is finite or countably infinite.

Countably Infinite: An infinite set is countably infinite if it is in 1-1 correspondence with the set of all natural numbers.

Cauchy Sequence: A sequence $\{x_n\}$ in a metric space is a Cauchy sequence if for every $\varepsilon > 0$, there is some positive integer n_0 such that $d(x_n, x_m) < \varepsilon$ for $n, m > n_0$.

Complete Metric Space: A metric space is complete if every Cauchy sequence converges in it.

Conditional Convergence of Series: A series is conditionally convergent if it is convergent, but not absolutely convergent.

Convergent Sequence: A sequence $\{x_n\}$ in a metric space (X, d) is convergent if for every $\varepsilon > 0$ there is a positive integer n_o such that $d(x_n, x) < \varepsilon$ for $n > n_o$.

Cover: A family $\{G_i\}$ of subsets of a metric space X is a cover for X if $\bigcup_i G_i = X$.

Compact: A set is compact if every open cover has a finite subcover.

Convex Function: A function f: $[a, b] \to R$ is convex if for $\lambda \in [0, 1]$ and $x, y \in [a, b]$,

$$f(\lambda x + (1 - \lambda)y) \leq \lambda f(x) + (1 - \lambda) f(y).$$

Divergent Sequence: A sequence which is not convergent is divergent.

Discontinuous Function: A function which is not continuous is discontinuous.

De Morgan's Laws: For any two sets A and B, $(A \cup B)^c = A^c \cap B^c$, $(A \cap B)^c = A^c \cup B^c$.

Equivalence Relation: A relation which is reflexive, symmetric and transitive.

Fundamental Theorem of Calculus: If f is Riemann integrable on $[a, b]$ and there exists a differentiable function F with $F' = f$ on $[a, b]$, then

$$\int_a^b f(x)dx = F(b) - F(a)$$

Greatest Lower Bound (Infimum): A real number c is the greatest lower bond of a subset of the real line if (i) it is a lower bound, (ii) if b is a lower bound other than c, then $c > b$.

Heine-Borel Theorem: Every closed and bounded subset of the real line is compact.

Intermediate Value Theorem: Every continuous function defined on a connected metric space attains all the intermediate values between any two values.

Infimum (Greatest Lower Bound): A number c is the infimum of a subset of the real line if (i) it is a lower bound, (ii) if b is a lower bound other than c, then $c > b$.

Inner Measure: Inner measure $m_c(E)$ of a set E on the real line is given by

$$m_c(E) = \sup_i l(F_i)$$

where $\{F_i\}$ is the collection of all closed sets contained in E.

Least Upper Bound (Supremum): A number c is the least upper bound of a subset of the real line if (i) c is an upper bound, (ii) if b is an upper bound other than c, then $c < b$.

Lower Bound: A real number b is called a lower bound of a nonempty subset S of the real line of $b \leq x$ for $x \varepsilon S$.

Limit Superior (Upper limit) and Limit Inferior (Lower limit): For a sequence $\{s_n\}$ of real numbers, let E be the set of all x for which $s_{n_k} \to x$ for some subsequence s_{n_k} of $\{s_n\}$.

Then

$$\limsup s_n = \sup E, \ \liminf s_n = \inf E.$$

Lagrange Mean-Value Theorem: If $f: [a, b] \to R$ is continuous on $[a, b]$ and differentiable on (a, b), then there exists a point $c \varepsilon (a, b)$ such that

$$\frac{f(b) - f(a)}{b - a} = f'(c).$$

Maximum Value Theorem: Every continuous function defined on a compact metric space attains its bounds.

Monotonically Increasing (Nondecreasing): A sequence $\{s_n\}$ of real numbers is monotonically increasing (nondecreasing) if $s_n \leq s_{n+1}$

for all $n = 1, 2, ...$. A function f is monotonically increasing (nondecreasing) if $f(x) \leq f(y)$ for $x < y$.

Monotonically Decreasing (Nonincreasing): A sequence $\{s_n\}$ of real numbers is monotonically decreasing (nonincreasing) if $s_n \geq s_{n+1}$ for all $n = 1, 2, ...$. A function f is monotonically decreasing (nonincreasing) if $f(x) \geq f(y)$ for $x < y$.

Measurable Set: A set is measurable if its inner- and outer- measures are equal.

Ordered Field: A field F is an ordered field if there exists a nonempty subset P of F such that $P \cup \{0\} \cup (-P) = F$.

Open Cover: A cover whose members are open sets.

Outer Measure: Outer measure $m_o(E)$ of a set E on the real line is given by

$$m_0(E) = \inf_i l(G_i)$$

where $\{G_i\}$ is the collection of all open sets containing E.

Partial Order Relation: A relation which is reflexive, anti-symmetric and transitive.

Power Series: A power series is a formal series of the form $\Sigma a_n x^n = a_o + a_1 x + a_2 x^2 + ... + a_n x^n + ...$

where $a_o, a_1, a_2, ... , a_n, ...$ are constants.

Relation: Relation is a set of ordered pairs.

Rolle's Theorem: If $f: [a, b] \to R$ is continuous on $[a, b]$, differentiable on (a, b) and $f(a) = f(b)$, then there exists a point $c \in (a, b)$ such that $f'(c) = 0$.

Reflexive: A relation R is reflexive if xRx for every x.

Radius of Convergence: Radius of convergence R of a power series $\Sigma a_n x^n$ is given by

$$\lim_{n \to \infty} |a_n|^{-1/n} \quad \text{or} \quad \lim_{n \to \infty} \left| \frac{a_n}{a_{n+1}} \right|.$$

Supremum (Least Upper Bound): A number c is the supremum of a subset of the real line if (i) c is an upper bound, (ii) if b is an upper bound other than c, then $c < b$.

Sequence: A sequence in a metric space is a mapping from the set of natural numbers into the space.

Subsequence: If $\{n_k\}$ is a sequence of positive integers such that $n_1 < n_2 < .....$, then $\{x_{n_k}\}$ is a subsequence of a given sequence $\{x_n\}$.

Symmetric: A relation R is symmetric if $xRy \Rightarrow yRx$.

Subcover: A sub-family of a cover which is also a cover.

Transitive: A relation R is transitive if xRy and $yRz \Rightarrow xRz$.

Upper Bound: A real number b is said to be an upper bound of a non-empty subset S of the real line if $x \leq b$ for all $x \in S$.

Uncountable: A set is uncountable if it is not countable.

Preface

This book is suitable as a text on 'Real Analysis' for BA/BSc (Pass and Hons) mathematics students of all Indian universities and Indian Institutes of Technology and second/third year undergraduate students of North America and Europe.

There are several excellent books available on the subject. However, there are not many titles published by Indian publishers or cheap editions by foreign publishers. Some of them are elementary, long and elaborate; some are advanced and too compact and become difficult for Indian undergraduate students to grasp. Therefore the authors felt the need of writing a book that was lucid and simple, neither too compact nor too elementary, that would contain the basics of a first course and also a number of examples to illustrate the subject.

The prerequisite for this book is only a course in Higher Secondary mathematics including calculus. However, efforts are being made to make the book self contained including a preliminary chapter on set theory. This course is prerequisite for advanced courses like Measure and Integration, Functional Analysis Topology and Complex Analysis which are taught at the MA/MSc level.

This course is taught as a compulsory course to BA/BSc students of all Indian universities and undergraduate students all over the world. The contents of this book have been taught by the first author at the Regional Engineering College, Rourkela, Utkal University, Bhubaneswar and Indian Institute of Technology, Kharagpur. During the preparation of this note the authors were greatly benefitted and have made use of works of several authors, a list of which appears in the references, The authors take this opportunity to acknowledge their indebtedness to them.

The manuscript was prepared when the first author was in the Department of Mathematics, IIT Kharagpur and the second author at Jiwaji University, Gwalior. The authors express their gratitude to the authorities of IIT Kharagpur and Jiwaji University, Gwalior for cooperation in preparing the manuscript and to IIT Kharagpur for financial support.

We have taken enough care to make the book free from errors, but despite our efforts, some mistakes may have survived. We will be grateful to the readers kind enough to point them out. Finally, we must express that the purpose will be best served if it benefits the students for whom it is meant.

S. NANDA
V.P. SAXENA

Contents

Preliminaries

SET THEORY

A set is a well-defined collection of certain distinguishable objects. The objects are called 'elements' or 'members' of the set.

We usually denote sets by capital letters such as A, B, C, ... and elements of a set by small letters such as a, b, c, ...To indicate that an element a is a member of a set A, we write

$$a \in A,$$

which is read 'a is an element of A', or 'a belongs to A' To indicate that the element a is not a member of the set A, we write

$$a \notin A,$$

which is read 'a is not an element of A' or 'a does not belong to A'. By a 'well-defined collection' is meant that given a set A and an object a, either $a \in A$, or $a \notin A$.

There are two standard notations available for designating a particular set. One way of describing a set is to list the elements of the set between braces. Thus the set of all positive integers that are less than 5 can be written as

$$\{1, 2, 3, 4\};$$

the set of all positive integers can be written as

$$\{1, 2, 3, ...\};$$

the set of all odd integers can be written as

$$\{\pm 1, \pm 3, \pm 5, ...\}$$

and the set of all the fourth roots of unity can be written as

$$\{1, i, -1, -i\}$$

With this notation an arbitrary set with six members may be written

$$\{a, b, c, d, e, f\}.$$

At times it is inconvenient or impossible to list all the elements of a set. In such cases, a set can often be described by specifying a property that the elements of the set have in common. For example, consider the set

$$A = \{1, 2, 3, 4, 5, 6\}$$

which has been described here by listing the elements. A is precisely the set of all positive integers that are less than 7 and this can be written as

$$A = \{x \mid x \text{ is a positive integer less than } 7\}$$

which is read A is the set of all x such that x is a positive integer less than 7. As another example consider the following expression

$$\{x \mid x \text{ is rational}\} \text{ or } \{x \mid x = \frac{m}{n},\ m \text{ and } n \text{ are integers, } n \text{ is positive}\}$$

which is read 'The set of all x, such that x is rational', denotes the set of all real numbers which can be written as the quotient of two integers. The four sets of numbers described at the beginning of this paragraph can be written either way:

$$\{1, 2, 3, 4\} = \{x \mid x \text{ is positive integer less than } 5\}$$

$$\{1, 2, 3, \ldots\} = \{x \mid x \text{ is positive integer}\},$$

$$\{1, i, -1, -i\} = \{x \mid x \text{ is a complex number and } x^4 = 1\}$$

$$\{\pm 1, \pm 3, \ldots\} = \{x \mid x \text{ is an odd integer}\}$$

Two sets A and B are equal if and only if they have the same elements. This is known as the *axiom of extension*. In this case we write

$$A = B.$$

If A and B are not equal, then we write

$$A \neq B.$$

For example, if

$$A = \{1, 2, 3,\}$$

and $\quad B = \{x \mid x \text{ is a positive integer and } x^2 < 16\}$

then $\quad A = B.$

Sometimes one sees sets that have no members. For example, the sets

$$\{x \mid x \text{ is rational and } x^2 = 2\},$$

$$\{x \mid x \text{ is real and } x^2 = -1\},$$

have no members. A set with no members is called an *empty set* or a *null set* and such a set is denoted by the symbol ϕ. A set is said to be *finite* if it is empty or consists of n elements for some positive integer n; otherwise it is said to be *infinite*.

More precisely, the existence of the empty set follows from the axiom of specification which says that, 'To every set A and to every condition $S(x)$ there corresponds a set B whose elements are exactly those elements x of A for which $S(x)$ holds'. By applying this axiom to any set A with the condition '$x \neq x$' we get the empty set

$$\{x \in A \mid x \neq x\}.$$

If every element of a set A is also an element of set B, then we say that A is a subset of B or is contained in B and write

$$A \subseteq B.$$

We sometimes express this by saying that B is a superset of A (or contains A). $A \subseteq B$ allows for the possibility that A and B might be equal. If A is a subset of B and A is not equal to B, then we say that A is a proper subset of B or is properly contained in B and this is written as

$$A \subset B.$$

We can also express this by saying that B is a proper superset of A or properly contains A. The existence of subsets follows from the axiom of specification. It is customary to agree that empty set $\varnothing$ is a subset of every set.

It is often convenient to have a symbol for logical implication. If p and q are statements, then $p \Rightarrow q$, means that p implies q, or that, if p is true, then q is true. Similarly $p \Leftrightarrow q$ means that p is equivalent to q, or that p is true if and only if q is true.

If A is any set, then

$$A \subseteq A$$

Let A, B and C be three sets. Then

$$A \subseteq B \text{ and } B \subseteq A \Rightarrow A = B,$$

$$A \subseteq B \text{ and } B \subseteq C \Rightarrow A \subseteq C.$$

Note that

$$A = B \Leftrightarrow A \subseteq B \text{ and } B \subseteq A.$$

If a set A is not a subset of B we write

$$A \not\subseteq B$$

Algebra of Sets

The intersection of two sets A and B is the set of all elements that belong to both A and B, it is denoted by the symbol

$$A \cap B.$$

If $A \cap B \neq \phi$, we say that A intersects B. If $A \cap B = \phi$, then we say that A does not intersect B or that A and B are disjoint. That $A \cap B$ is a set follows from the axiom of specification if we write

$$A \cap B = \{x \in A \mid x \in B\} \quad \text{or} \quad A \cap B = \{x \in B \mid x \in A\}.$$

The intersection of any collection of sets is the set of all elements that belong to every one of the sets in the collection.
The operation of forming intersections is commutative and associative:

$$A \cap B = B \cap A$$

and

$$A \cap (B \cap C) = (A \cap B) \cap C$$

It also has the property that

$$A \cap A = A, \; A \cap \phi = \phi.$$

The union of two sets A and B is the set of all elements that belong to either A or B or both, it is denoted by the symbol

$$A \cup B$$

The above definition can also be expressed as

$$A \cup B = \{x \mid x \in A \text{ or } x \in B\}$$

The operation of forming unions is also commutative and associative:

$$A \cup B = B \cup A$$

and

$$A \cup (B \cup C) = (A \cup B) \cup C$$

It also has the properties that

$$A \cup A = A, \; A \cup \phi = A$$

The union of any collection of sets is the set of all elements that belong to one or more of the sets in the collection. That it is a set follows from the axiom of union: For any collection of sets there is a set that contains all the elements that belong to at least one set of the given collection.

We have now defined two of the fundamental operations on sets and we shall now see how they are related to one another. The facts here are given by the distributive laws:

$$A \cap (B \cup C) = (A \cap B) \cup (A \cap C)$$
$$A \cup (B \cap C) = (A \cup B) \cap (A \cup C)$$

The next of our major operations on sets is the formation of complements.

Given any set A, it may be necessary to consider those elements that are not in A. Usually, however, there are elements not in A that are extraneous to the problem under consideration. For example, let A be the set of all positive integers. Clearly, if x is a negative integer, then $x \notin A$ However, if x is a cabbage then x is also not in A since cabbages are not positive integers. If we are concerned with a problem about integers then 'an element x not in A' is really meant 'an integer x that is not in A '. Cabbages are extraneous to the problem.

One way of eliminating extraneous elements is to specify in advance that all elements in the problem under consideration come from some fixed set U, called the *universal set*. For example, in a problem concerned with real numbers, we might agree to take the universal set U to be the set of real numbers and by 'an element x that is not in A' we must mean 'a real number x that is not in A,. In many problems the universal set is clear from the nature of the problem and is not explicitly stated.

If U is the universal set and A is a subset of U, then the set of all elements in U that are not members of A is called the complement of A; it is denoted by the symbol

$$A^c$$

The operation of forming complements has the following obvious properties:

$$(A^c)^c = A, \; U^c = \phi, \; \phi^c = U$$

$$A \cup A^c = U, \; A \cap A^c = \phi.$$

The following two important results, called *De Morgan's laws* , give relationship of complementation with union and intersection.

If A and B are any two sets, then

$$(A \cup B)^c = A^c \cap B^c$$

$$(A \cap B)^c = A^c \cup B^c.$$

We now prove the first De Morgan's law.

Let $x \in (A \cup B)^c$

$$\Leftrightarrow \quad x \notin A \cup B$$

$$\Leftrightarrow \quad x \notin A \quad \text{and} \quad x \notin B$$

$$\Leftrightarrow \qquad x \in A^c \quad \text{and} \quad x \in B^c$$

$$\Leftrightarrow \qquad x \in A^c \cap B^c.$$

Hence

$$(A \cup B)^c = A^c \cap B^c.$$

We now introduce the difference and symmetric difference of two sets. If A and B are two sets, we define

$$A - B = \{x \in A \mid x \notin B\}$$

That this is a set follows from the axiom of specification. It is easy to see that

$$A - B = A \cap B^c$$

We define the symmetric difference of two sets A and B as

$$A \Delta B = (A - B) \cup (B - A).$$

This operation is commutative and associative, i.e.

$$A \Delta B = B \Delta A$$

$$A \Delta (B \Delta C) = (A \Delta B) \Delta C.$$

Also

$$A \Delta \phi = A, \quad A \Delta A = \phi.$$

It can be checked easily that

$$A \Delta B = (A \cup B) - (A \cap B).$$

We have been considering the subsets of a given universal set U; do those subsets themselves constitute a set? The following axiom guarantees that the answer is yes.

Axiom of Powers: For each set there exists a collection of sets that contains among its elements all the subsets of the given set.

In other words, if U is a set, then there exists a set $\wp(U)$ such that if $X \subset U$, then $X \in \wp(U)$.

The set $\wp$ described above may be larger than wanted; it may contain elements other than the subsets of U. This is easy to remedy; just apply the axiom of specification to form the set $(\{X \in \wp(U) : X \subset U\}$. Since for every X, a necessary and sufficient condition that X belongs to this set is that X be a subset of U, it follows that if we change notation and call this set $\wp(U)$ again, then

$$\wp(U) = \{X : X \subset U\}$$

The set $\wp(U)$ is called the power set of U; the axiom of extension guarantees its uniqueness.

Because the set $\wp(U)$ is very big in comparison with U, it is not easy to give examples. If $U = \phi$, then

$$\wp(U) = \{\phi\},\ \wp(\{a\}) = \{\ \phi,\ \{a\}\}\ \text{ and}$$

$$\wp(\{a,\ b\}) = \{\phi,\ \{a\},\ \{b\},\ \{a,\ b\}\}$$

The power set of triple has eight elements. The reader can probably guess (and is hereby challenged to prove) the generalisation that includes all these statements: the power set of a finite set with n elements has 2^n elements.

RELATIONS

In many situations we are interested in paired data. An ordered pair (a, b) is a listing of two objects, a and b in a definite order. The element a is called the first entry and the element b is called the second entry of the ordered pair. Two ordered pairs are said to be equal if they list the same objects in the same order. Thus, for (a, b) and (c, d) to be equal, we must have

$$a = c \text{ and } b = d$$

Thus two ordered pairs (a, b) and (b, a), though involving the same elements, convey different information because of their differing order unless $a = b$.

We are now in a position to define the Cartesian product of sets. If A and B are two sets, the Cartesian product of A and B is the set of all ordered pairs (a, b) where $a \in A$ and $b \in B$, which is expressed by:

$$A \times B$$

Thus

$$A \times B = \{(a,\ b)\ |\ a \in A,\ b \in B\}$$

Example: If $A = \{\ a,\ b,\ c\}$

and $B = \{1,\ 2\}$

then

$$A \times B = \{a,\ 1),\ (a,\ 2),\ (b,\ 1),\ (b,\ 2),\ (c,\ 1),(c,\ 2)\}.$$

It is possible to define the Cartesian product of more than two sets. For example, if $A_1, A_2, \ldots A_n$, are n sets, where n is a positive integer, then the Cartesian product

$$A_1 \times A_2 \times \ldots \times A_n$$

is defined by

$$A_1 \times A_2 \times \dots A_n = \{(a_1, \dots, a_n) \mid a_1 \in A_1 \dots, a_n \in A_n\}$$

A *relation* is any set of ordered pairs. Thus a relation is any set which is a subset of the Cartesian product of two sets. Let X and Y be any two sets (which may be equal, but need not be) and let R be a relation from X into Y. Then $R \subset X \times Y$. We define the domain of R to be the set, dom $R = \{ x \in X : (x, y) \in R$ for some $y \in Y \}$ and the range of R to be the set, range $R = \{ y \in Y : (x, y) \in R$ for some $x \in X\}$. The symbol R^{-1} denotes the inverse of R: $R^{-1} = \{y, x): (x, y) \in R\}$.

There are two types of relations which often arise in mathematics: *equivalence relations* and *order relations*. We first discuss equivalence relations.

Let X be any non empty set. An equivalence relation on X is any relation $\sim \subset X \times X$ such that, for all x, y, z, in X we have

(i) $x \sim x$ (reflexive);

(ii) $x \sim y \Rightarrow y \sim x$ (symmetric);

(iii) $x \sim y$ and $y \sim z \Rightarrow x \sim z$ (transitive)

An equivalence relation on X has an intimate connection with decompositions of X into nonempty subsets which fill it out and have no elements in common with one another.

A partition of X is a disjoint class $\{ X_i \}$ of nonempty subsets of X whose union is the full set X itself.

In other words, a partition of X is the result of splitting it, or subdividing it into nonempty subsets in such a way that each element of X belongs to one and only one of the given subsets.

We now show that each partition of X has associated with it a natural equivalence relation in X and, conversely, a given equivalence relation in X determines a natural partition of X.

Let X be a nonempty set. Assume that a partition of X is given . We now associate with this partition a relation '$\sim$' in X. We say that $x \sim y$ if x and y belong to the same partition set. It is easy to see that $\sim$ defines an equivalence relation in X.

Conversely, let $\sim$ be an equivalence relation in X. Let x be an arbitrary element of X. The set defined by $[x] = \{ y : y \sim x\}$, which is the set of all elements of x which are equivalent to x, is called the equivalence set of x. We show that the class of all distinct equivalence sets forms a partition of X. By reflexivity $x \in [x]$ for each $x \in X$, so each equivalence set is nonempty and their union is X. It remains to show that any two equivalence sets $[x_1]$ and $[x_2]$ are either disjoint or identical. Suppose that $[x_1]$ and $[x_2]$ are not disjoint. To be precise, let $z \in [x_1] \cap [x_2]$. Now $z \sim x_1$; and $z \sim x_2$, and by symmetry $x_1 \sim z$. Let $y \in [x_1]$, so that $y \sim x_1$; now, since $y \sim x_1$ and $x_1 \sim z$, transitivity shows that $y \sim z$. Since $y \sim z$ and $z \sim x_2$, again transitivity shows that $y \sim x_2$, so $y \in [x_2]$. Since y is an arbitrary element of $[x_1]$, this shows that

$$[x_1] \subseteq [x_2]$$

The same reasoning shows that $[x_2] \subseteq [x_1]$

Thus, $[x_1] = [x_2]$ and this completes the proof.

The following is a simple example of equivalence relation.

Example 2

Let Z be the set of all integers. For a, $b \in Z$ we define $a \sim b \Leftrightarrow a = b$.

Then it is evident that ~ defines an equivalence relation in the set Z and each equivalence class consists of precisely one integer.

We now turn to the concept of *order relations*.

Let X be any nonempty set. A partial order relation (por) in X is a relation $\leq \subset X \times X$ such that for all $x, y, z \in X$ we have

(i) $x \leq x$ (reflexive);
(ii) $x \leq y$ and $y \leq x \Rightarrow x = y$ (antisymmetric);
(iii) $x \leq y$ and $y \leq z \Rightarrow x \leq z$ (transitive).

We sometimes write $x \leq y$ in the equivalent form $y \geq x$. A nonempty set X with a partial order relation defined on it is called a partially ordered set (poset). We now give three important and simple examples of partially ordered sets.

Example 3

Let X be the set of all positive integers, and $x \leq y$ mean that x divides y.

Example 4

Let X be the set of all real numbers, and $x \leq y$ mean that x is less than or equal to y in the usual sense.

Example 5

Let X be the family of all subsets of some universal set U and let $A \leq B$ mean $A \subseteq B$.

Two elements x and y in a poset are called *comparable* if either $x \leq y$ or $y \leq x$. Note that there may be pairs of elements in a poset which are not comparable. In Example 3, for example, the integers 3 and 7 are not comparable because neither divides the other; and in Example 5 if U has more than one element, there are always two subsets of U, neither of which is a subset of the other.

Some *partial order relations* possess a fourth property, in addition to the three required by the definition:

(iv) any two elements are comparable. A por with property (iv) is called a total (or *linear*) *order relation* and a set with a total (or linear) order relation defined on it is called a *totally ordered set* or a *linearly ordered set* or a *chain.*

FUNCTIONS

Let X and Y be any two sets (which may be equal but need not be). A function from X into Y is a rule 'f' which assigns to each element $x \in X$ a unique element $y \in Y$. The y which corresponds in this way to a given x is usually written $f(x)$, and is called the image of x under the rule 'f', or the value of 'f' at the element x. A function 'f' can also be viewed as a relation, that is, a subset of $X \times Y$ in which $(x, y) \in f$ and $(x, z) \in f$ implies that $y = z$.

A function f from X into Y is often called a *mapping*, or a *map*, or a *transformation* from X, into Y. The set X is called the domain of f, written as 'dom f', and the set of all values of f for all x in X is called its range and is written as 'rng f'. A function whose range consists of just one element is called a constant function.

If f is a function from X into Y it is denoted by the symbol

$$f: X \to Y$$

If it is clear from the context what the sets X and Y are, or if there is no need of specifying them explicitly, it is common practice to identify the function $f: X \to Y$ by the rule f without mentioning the sets X and Y. If two definite sets X and Y are under discussion, then a mapping of X into Y is sometimes designated by

$$x \to y$$

where y is the image of x. For example, the function $f: R \to R$, where R denotes the set of all real numbers defined by

$$f(x) = x^2$$

can be designated by

$$x \to x^2.$$

A function f is called an *extension* of a function g (and g is called a *restriction* of f) if the domain of f contains the domain of g and $f(x) = g(x)$ for each x in the domain of g.

If $f : X \to Y$, let

$$G_f = \{(x, y) \in X \times Y : y = f(x)\}.$$

This set G_f is called the graph of f. the graph G_f is related to the function f in the sense that either one of them completely determines the other.

Consider a map $f: X \to Y$. Then clearly range of f is a subset of Y, that is range $f \subseteq Y$. If $f: X \to Y$ is such that range of f equals Y, then we call f surjective, or a mapping of X onto Y. If any two distinct elements in X have distinct images under f, then we call f injective, or a 1 – 1 mapping from X into Y. If f is both injective and surjective, then we call it bijective. If $f : X \to Y$ is bijective, then we can define its inverse mapping $f^{-1} : Y \to X$ as follows : for each $y \in Y$ there exists a unique element $x \in X$ such that $f(x) = y$ since f is 1–1 and onto; we than define x to be $f^{-1}(y)$.

Let X be a nonempty set. The identity mapping on X is the mapping of X onto itself which leaves each element of X fixed.

Suppose that $f : X \to Y$ and $g : Y \to Z$ be any two mappings. The product or composition of these two mappings, denoted by $gf : X \to Z$, is defined by

$$(gf)(x) = g(f(x))$$

We observe that the two mappings involved here are not arbitrary, for the set Y which contains the range of the first equals the domain of the second. More generally, the product of two mappings is meaningful whenever the range of the first is contained in the domain of the second.

Example 6

$f : R \to R$ be defined by

$$y = f(x) = x^2 + 1$$

and $g : R \to [-1, 1]$ be defined by

$$g(x) = \text{Sin } x.$$

Then $gf : R \to [-1, 1]$ is defined by

$$(gf)(x) = g(f(x)) = g(x^2+1)$$
$$= \text{Sin}(x^2 + 1).$$

Consider an arbitrary mapping $f: X \to Y$. If $A \subseteq X$, then its image $f(A)$ is the subset of Y defined by

$$f(A) = \{f(x) \mid x \in A\}$$

Similarly if $B \subseteq Y$, then its inverse image $f^{-1}(B)$ is the subset of X defined by

$$f^{-1}(B) = \{x \mid f(x) \in B\}.$$

The first mapping has the following properties:

$$f(\phi) = \phi; f(X) \subseteq Y;$$
$$A_1 \subseteq A_2 \Rightarrow f(A_1) \subseteq, f(A_2)$$
$$f(\bigcup_{\iota \in I} A_i) = \bigcup_{\iota \in I} f(A_i)$$
$$f(\bigcap_{\iota \in I} A_i) \subseteq \bigcap_{\iota \in I} f(A_i)$$

The second one is much better behaved:

$$f^{-1}(\phi) = \phi; f^{-1}(Y) = X,$$
$$B_1 \subseteq B_2 \Rightarrow, f^{-1}(B_1) \subseteq f^{-1}(B_2)$$
$$f^{-1}(\bigcup_{\iota \in I} B_i) = \bigcup_{\iota \in I} f^{-1}(B_i)$$
$$f^{-1}(\bigcap_{\iota \in I} B_i) = \bigcap_{\iota \in I} f^{-1}(B^{-i})$$

ELEMENTARY LOGIC

We use sentence to communicate ideas. Statements are sentences which have the property of being either true or false. Statements can be illustrated by examples like 'this pen is black', 'this man is old', 'three plus seven equals ten'. Since statements can be false, 'three plus seven equals nine' is also an example of a statement. It may be noted that the truth or falsity of a statement often depends upon which elements of the universe of discourse is being considered, e.g., which pen, or which man.

Let A and B denote any two statements. The statement 'A and B' is called the conjunction of A and B and is written as $A \wedge B$. The disjunction of A and B is the statement 'A or B' and is written as $A \vee B$. If T and F denote 'true' and 'false' respectively, we can prepare truth tables for conjunction and disjunction which read as follows:

Conjunction

A	B	$A \wedge B$
T	T	T
T	F	F
F	T	F
F	F	F

Disjunction

A	B	$A \vee B$
T	T	T
T	F	T
F	T	T
F	F	F

To deny a statement *A* is to affirm another statement known as, the negation of A and is written as 'not *A*' or 7A. The truth table for negation is as follows:

Negation

A	$7A$
T	F
F	T

The implication is a statement of the form 'if *A*, then *B*'. This is symbolised as $A \Rightarrow B$ and is read as '*A* implies *B*'. The statement *A* associated with the word 'if', is called the *hypothesis*, and the statement *B* associated with the word 'then' is called the *conclusion*. An implication is false only if the hypothesis is true and the conclusion is false. The truth table for implication is constructed as follows:

Implication

A	B	$A \Rightarrow B$
T	T	T
T	F	F
F	T	T
F	F	T

If we write, 'if *B*, then *A*' , i.e, '$B \Rightarrow A$', then we have a new sentence, called the converse of the original one. 'If not A, then not *B*', i.e., $7A \Rightarrow 7B$, is a new implication, called the inverse of the original one. The implication 'if *not B*, then not *A*' i.e., '$7B \Rightarrow 7A$' is termed contrapositive to the original one. We construct the following truth table by connecting negation and implication.

From this table one shoule note that the validity of the original proposition and the contrapositive always agree, in the sense that if one is true, so is the other; or if one is false, so is the other, Also observe that the validity of converse and inverse always agree.

		Original	Converse			*Inverse*	*Contrapositive*
A	B	$A \Rightarrow B$	$B \Rightarrow A$	$7A$	$7B$	$7A \Rightarrow 7B$	$7B \Rightarrow 7A$
T	T	T	T	F	F	T	T
T	F	F	T	F	T	T	F
F	T	T	F	T	F	F	T
F	F	T	T	T	T	T	T

The equivalence of two statements A and B is defined by the conjunction $A \Rightarrow B$ and $B \Rightarrow A$. This statement is written 'A if and only if B' or, 'A iff B' or, '$A \Leftrightarrow B$' and read 'A is equivalnet to B'. This helps us in expressing a theorem with a valid converse in a single sentence.

It is easy to write down the truth table for equivalence from the truth tables of conjunction and implication. It should be noted that if $A \Leftrightarrow B$ is a true statement, then A and B have the same truth values, i.e., either both are true, or both are false.

Equivalence

A	B	$A \Rightarrow B$	$B \Rightarrow A$	$A \Leftrightarrow B$
T	T	T	T	T
T	F	F	T	F
F	T	T	F	F
F	F	T	T	T

A statement that is always true is called a *tautology*; a statement that is always false is an *absurdity*; one that is sometimes true and sometimes false is a *contingency*. Note that whenever any statement A is an absurdity, its negation $7A$ is a tautology. If A and B are any two statements, then $A \Rightarrow (A \wedge B)$ is a contingency; $(A \wedge B) \Rightarrow A$, $(A \wedge B) \Leftrightarrow (B \wedge A)$ and $A \vee 7A$ are tautologies; where as $A \wedge 7A$ is an absurdity.

The truth table for the proposition $(B \wedge 7A) \Rightarrow A$ can be given as follows:

A	B	$7A$	$B \wedge 7A$	$B \wedge 7A \Rightarrow A$
F	F	T	F	T
F	T	T	T	F
T	F	F	F	T
T	T	F	F	T

Using truth tables, a number of useful equivalences involving negation, disjunction and conjunction of propositions can be derived. Some of these equivalences, referred to as *propositional laws*, are shown below where A, B and C denote arbitrary propositions, and 1 and o denote tautology and contradiction respectively.

Propositional Laws

Commutative laws:	$(A \vee B) \Leftrightarrow B \vee A$, $(A \wedge B) \Leftrightarrow (B \wedge A)$.
Associative laws:	$(A \vee (B \vee C)) \Leftrightarrow (A \wedge (B \wedge C)) \Leftrightarrow ((A \vee B) \vee C\ ((A \wedge B) \wedge C)$.
Distributive laws:	$(A \wedge (B \vee C)) \Leftrightarrow (A \vee (B \wedge C)) \Leftrightarrow ((A \wedge B) \vee (A \wedge C)$.
Identity laws:	$(A \vee 0) \Leftrightarrow (0 \vee A) \Leftrightarrow A$; $(A \wedge 1) \Leftrightarrow (1 \wedge A) \Leftrightarrow A$.
Negation laws:	$(A \vee 7A) \Leftrightarrow 1$; $(A \wedge 7A) \Leftrightarrow 0$.
Idempotent laws:	$(A \vee A)\ A \Leftrightarrow$; $(A \wedge A) \Leftrightarrow A$..
Null laws:	$(A \vee 1) \Leftrightarrow 1$; $(A \wedge 0) \Leftrightarrow 0$.
Absorption laws:	$(A \vee (A \wedge B)) \Leftrightarrow A$; $(A \wedge (A \vee B)) \Leftrightarrow A$.
De Morgan's laws:	$(7(A \vee B)) \Leftrightarrow (7(A \vee B)) \Leftrightarrow ((7A) \wedge (7B))$; $((7A) \vee (7B))$.

We now mention some of the important implications and equivalences:

Detachment:	$((A \Rightarrow B) \wedge A) \Rightarrow B$.
Disjunctive addition:	$A \Rightarrow (A \vee B)$.
Conjunctive simplification:	$(A \wedge B) \Rightarrow A$.
Chain rule:	$((A \Rightarrow B) \wedge (B \Rightarrow C)) \Rightarrow (A \Rightarrow C)$.
Disjunctive simplification:	$((A \vee B) \wedge (7A)) \Rightarrow B$.
Conditional rules:	$(A \Rightarrow B) \Leftrightarrow ((7A) \vee B)$. $(A \Rightarrow B) \Leftrightarrow 7B \Rightarrow 7A$.
Equivalence rules:	$(A \Rightarrow B) \Leftrightarrow ((A \Rightarrow B) \vee (B \Rightarrow A))$; $(A \Leftrightarrow B) \Leftrightarrow ((A \wedge B) \vee ((7A) \wedge (7B))$; $(7\ (A \Leftrightarrow B)) \Leftrightarrow (A \Leftrightarrow (7B))$.
Implication rules:	$(A \Rightarrow B) \Rightarrow ((A \wedge C) \Rightarrow (B \wedge C))$; $(A \Rightarrow B) \Rightarrow ((A \vee C) \Rightarrow (B \vee C)$.
Contrapositive inference:	$((A \Rightarrow B) \wedge (7B)) \Rightarrow (7A)$.

EXERCISES

1. Mark right or wrong

(i) $\{-1, 0, 1\} = \{x \in Z : |x| \le 1\}$.

(ii) $\left\{\frac{m}{n} : m \text{ and } n \text{ are integers}\right\} = \{x : x \text{ is rational}\}$

(iii) $\{1, -1\} = \{x: x^2 = 1\}$.

(iv) $A \subseteq B, B \subseteq C \Rightarrow A \subseteq C$.

(v) $A \cap B = B \cap A$.

2. State whether true or false? (U stands for the universal set)

(i) $U \cap \phi = \phi$

(ii) $U \cup V = U \cap V$ for $V \subset U$.

(iii) $A^c \cup A \neq U$.

(iv) $(A^c \cup A) \cap (A^c \cap A) = U$.

(v) $(A^c \cup B^c) \cup (A \cup B) = U$, where $A, B \subseteq U$.

(vi) $A \times B = B \times A \Rightarrow A = B$.

3. State which of the following are correct:

(i) $\{A, B\}$ is a partition of U where $A \cup B = U$ for all $A, B \subset U$.

(ii) $f : R \to R$ is a mapping where $f(x) = \sqrt{x}$.

4. Write equivalent symbolic forms of the following sets:

(i) $\{1, 3, 5, \ldots, 2n+1\}$

(ii) $\{i, -1, -i, 1\}$

(iii) $\{1/2, 2/3, 3/4, 4/5\}$

(iv) $\{0.1, 0.01, 0.001\}$

(v) $\{\pm 1, \pm 2, \pm 3\}$.

5. Write all the subsets of the following:

(i) $\{0, 1\}$

(ii) $\{x : 1 < x^2 < 16, x \text{ is a positive integer}\}$

(iii) $\{[x] : 0 < x < 3\}$.

(iv) $\left\{\frac{x}{x+1} : x = \pm\frac{1}{2}\right\}$

(v) $\{x^n : x = \pm 1 \text{ and } n = 1, 2, 3\}$.

6. State whether the following statements are true or false.

(i) $(A \cup B) \cup C = A \cup (B \cup C)$.

(ii) $(A \cup B) - A = B$.

(iii) $(A \cup B) - (A \cap B) = A \cup (B - A)$.

(iv) $(A \cap (B \cup C) = (A \cap B) \cup (A \cap C)$.

(v) $(A^c \cap B^c)^c = A \cup B$.

7. Write $A \times A$ where $A = \{-1, 0, 1\}$.

8. Partition X in a disjoint class of subsets where $X = \{0, \pm 1, \pm 2\}$.

9. Define a suitable function, f, which maps R into N. Is the inverse mapping possible?

10. Find R' where the function f, defined by

$$f(x) = |x|$$

maps R onto R'.

11. Let $f: X \to Y$ be an arbitrary mapping. Define a relation in X as follows: $x_1 \sim x_2$ means that $f(x_1) = f(x_2)$. Show that this is an equivalence relation and describe the equivalence sets.

12. In the set R of all real numbers, let $x \sim y$ mean that $x - y$ is an integer. Show that this is an equivalence relation and describe the equivalence sets.

13. Let m be a fixed positive integer. Two integers a and b are said to be congruent modulo m—symbolised by $a \equiv b \pmod{m}$, if $a - b$ is exactly divisible by m, i.e., if $a-b$ is an integral multiple of m. Show that this is an equivalence relation, describe the equivalence sets.

14. Decide which ones of the four properties of reflexivity, anti-symmetry, symmetry, and transitivity, are true for each of the following relations in the set of positive integers: $m \le n$, $m < n$, m divides n, $m = n$.

15. Let X be a nonempty set. A relation $\sim$ in X is called *circular* if $x \sim y$ and $y \sim z \Rightarrow z \sim x$, and *triangular* if $x \sim y$ and $x \sim z \Rightarrow y \sim z$. Prove that a relation in X is an equivalence relation $\Leftrightarrow$ it is reflexive and circular $\Leftrightarrow$ it is reflexive and triangular.

16. Give an example each of a relation which is (a) reflexive but not symmetric or transitive; (b) symmetric but not reflexive or transitive; (c) transitive but not reflexive or symmetric; (d) reflexive and transitive but not symmetric; (e) symmetric and transitive but not reflexive.

17. Let X be a nonempty set, and let f and g be one-to-one mappings of X onto itself. Show that fg is also a one-to-one mapping of X onto itself and that $(fg)^{-1} = g^{-1} f^{-1}$.

18. Let f, g and h be any three mappings of a nonempty set X into itself, and show that multiplication of mappings is associative in the sense that $f(gh) = (fg)h$.

19. Let X and Y be nonempty sets and f a mapping of X into Y. Show the following:

(a) f is one-to-one $\Leftrightarrow$ there exists a mapping g of Y into X such that $gf = id_X$;

(b) f is onto $\Leftrightarrow$ there exists a mapping h of Y into X such that $fh = id_Y$.

20. Let X be a nonempty set and f a mapping of X into itself. Show that f is one-to-one onto $\Leftrightarrow$there exists a mapping g of X into itself such that $fg = gf = id_X$.

21. Construct the truth table of the following propositions:

(a) $(A \wedge B) \vee (\neg C)$,

(b) $(A \vee B) \wedge ((\neg A) \vee (\neg C)$,

(c) $(\neg A) \vee B$,

(d) $A \wedge (\neg B)$,

(e) $\neg((\neg A) \wedge (\neg B))$,

(f) $(\neg A) \vee (B \wedge (\neg C))$, and

(g) $(A \wedge (\neg B)) \vee (C \wedge B)$.

22. Construct the truth tables of the following propositions:

(a) $(A \wedge (\neg B)) \Rightarrow ((\neg A) \wedge (\neg C))$,

(b) $((A \vee B) \Rightarrow (B \wedge C)) \Rightarrow (A \wedge (\neg C)$,

(c) $(A \wedge B) \Rightarrow A$,

(d) $(B \wedge (\neg A)) \Rightarrow A$,

(e) $A \Rightarrow (A \vee B))$,

(f) $((A \Rightarrow B) \wedge A) \Rightarrow B$.

23. Construct the truth tables of the following propositions:

(a) $(A \wedge (\neg B)) \Leftrightarrow ((\neg A) \vee (\neg C))$,

(b) $((A \vee B) \Leftrightarrow (B \wedge C)) \Leftrightarrow (A \wedge (\neg C))$,

(c) $(A \Rightarrow B) \Leftrightarrow ((\neg A) \vee B)$,

(d) $(A \Leftrightarrow B) \Leftrightarrow ((A \wedge B) \vee ((\neg A) \wedge (\neg B)$,

(e) $((A \wedge B) \vee \neg C) \Leftrightarrow A$.

1 Real and Complex Number Systems

1.1 INTEGERS AND RATIONALS

The oldest and the most basic mathematical system is the set N of all positive integers. Kronecker once said 'God made the positive integers, all else is due to man'. However, it was shown by the Italian mathematician G. Peano (1858 - 1932) that the positive integers can even more simply be defined by a set of axioms, known as Peano axioms. To describe these axioms we first need to define the so-called Peano's successor function. This function, denoted by s, assigns to every positive integer the next larger one. Thus

$$s : N \to N$$

is defined by

$$s(n) = n + 1.$$

In terms of the function s, a number of basic properties can be formulated which characterise the set N. These properties can be summarised as follows:

(a) $1 \in N$

(b) $n \in N \Rightarrow s(n) \in N$

(c) $\exists$ no $n \in N$ such that $s(n) = 1$

(d) $s(n) = s(m) \Rightarrow n = m$

(e) if $A \subset N$ such that $1 \varepsilon A$ and $s(n) \varepsilon A$ whenever $n \varepsilon A$, then $A = N$.

Axiom (e) leads to one of the most important principles of mathematics, known as the *principle of mathematical induction*. This is a powerful tool in proving theorems. We shall now illustrate some examples of proof by induction.

Examples

Prove the following statements by induction on n:

(a) $\sum_{i=1}^{n} i = \frac{1}{2} n(n+1) \; (n \in N)$

(b) $\sum_{i=1}^{n} i^2 = \frac{1}{6} n(n+1)(2n+1) \; (n \in N)$

(c) $\sum_{i=1}^{n} i^3 = \left[\frac{1}{2} n(n+1)\right]^2 \; (n \in N)$

We only prove (a), the others are left to the reader as exercises. For $n = 1$, the theorem is true, because in this case

$$\sum_{i=1}^{n} i = \frac{1}{2}.1.2 = 1$$

The induction hypothesis is that of some $k \in N$,

$$\sum_{i=1}^{k} i = \frac{1}{2} k(k+1)$$

Now

$$\sum_{i=1}^{k+1} i = \left(\sum_{i=1}^{k} i\right) + (k+1)$$

$$= \left[\frac{1}{2} k(k+1)\right] + (k+1) = \frac{1}{2}(k+1)(k+2)$$

which proves the theorem for $n = k + 1$. This completes the proof.

Let Z denote the set of all integers (positive, negative and zero). Let D be the set of all ordered pairs (a,b) where $a, b \varepsilon Z$ and $b \neq 0$. One can write a/b for (a,b) and call it a fraction. In D define a relation $\sim$ as follows : $(a,b) \sim (c,d)$ if and only if $ad = bc$. It is easy to see that $\sim$ is infact an equivalence relation in D. Hence $\sim$ divides D into mutually disjoint equivalence classes. Let $[a,b]$ be the equivalence class in D of (a,b) and let Q be the set of all such equivalence classes $[a, b]$ where $a, b \in Z$ and $b \neq 0$. This set Q is infact the set of all rational numbers. In other words, a rational number is an equivalence class $[a, b]$ of an ordered pair (a, b) where $a, b \in Z$ and $b \neq 0$. *One can write* a/b for a rational number (a, b), but there should not be confusion between a rational number a/b and a fraction a/b. One should note that, as ordered pairs, (1, 2) and (2, 4) or, as fractions, $\frac{1}{2}$ and 2/4, are distinct, but as rational numbers both are identical, because both are in the same equivalence class. Further, notice that for $x \neq 0$, $y \neq 0$ in Z, $[ax, x] = [ay, y]$ since $(ax)y = x(ay)$. Let us denote $[ax, x]$ by $[a, 1]$ and identify the integer a with the rational number $[a, 1]$. In this sense the set Z can be viewed as a subset of Q.

1.2 REAL NUMBERS

We shall now introduce the concept of real numbers. If we take the length of any side of a square as a unit, then the length of any diagonal of that square cannot be expressed by a rational number. In other words, there is no rational number whose square is 2. To prove this, let x denote the diagonal. Then

$$x^2 = 1^2 + 1^2 = 2.$$

Suppose that $x = \frac{a}{b}$ where a and b are integers, coprime to each other and $b \neq 0$. Then

$$2 = x^2 = \frac{a^2}{b^2}$$

Hence

$$a^2 = 2b^2$$

The square of an odd integer is not divisible by 2, thus a is even, say $a = 2c$. Hence

$$4c^2 = 2b^2 \text{ or } b^2 = 2c^2$$

Thus b is even, say $b = 2d$. Therefore a and b cannot be coprime and this contradiction proves the assertion that $x = \frac{a}{b}$ is impossible.

This demonstrates that there is no rational number whose square is equal to 2. Therefore we can divide the set of all positive rational numbers into two classes. Let U be the class of rationals whose squares are greater than 2 and L the class of these whose squares are less than 2.

If $l \in L$, then we can find another element of L which is greater than l, in other words, L has no greatest member. For if $l \in L$, then we can choose another rational number h, $0 < h < 1$, such that $p = l + h$ also belongs to L.

Thus

$$p^2 = l^2 + h(2l + h) < l^2 + h(2l + 1) < l^2 + 2 - l^2 = 2$$

provided $$h < (2 - l^2)/(2l + 1).$$

In a similar vein it can be shown that U, the set of all positive rational numbers whose squares are greater than 2 has no least member.

It is clear that every element of L is less than every element of U. Also it is possible to find an element of L and an element of U whose difference is as small as we please. For, let $\varepsilon > 0$ be arbitrarily small. Let $l \varepsilon L$ and $r \varepsilon U$ be such that

$$2 - l^2 < \frac{\varepsilon}{2} \text{ and } r^2 - 2 < \frac{\varepsilon}{2}$$

Then

$$r^2 - l^2 = (r^2 - 2) + (2 - l^2) < \frac{\varepsilon}{2} + \frac{\varepsilon}{2} = \varepsilon$$

Now since $l > 1$ and $r > 1$ we must have

$$r - l = \frac{r^2 - l^2}{r + l} < \varepsilon$$

Let α denote the union of the set of all nonpositive rational numbers and L.

Now if $a \in \alpha$, then clearly any rational number less than a is also in α.

Also α has no greatest member.

The set α of rational numbers is said to define the irrational number whose square is 2 and this irrational number is denoted by the symbol $\sqrt{2}$. α is called a *section* of rational numbers.

The motivation of this example is used to define any (and hence all) irrational numbers. This was first demonstrated by the famous mathematician Dedekind and the method is known as Dedekind *cut* or Dedekind *section*. We shall discuss Dedekind's theory in the following section which defines irrational numbers. All rational and irrational numbers constitute the real number system. We write R to denote the set of all real numbers.

1.3 DEDEKIND SECTION

Dedekind's definition of an irrational number is based on a set α of rational numbers, called a *section*, which satisfies the following axioms:

(i) α is nonempty and $\alpha \neq Q$; in otherwords α contains at least one rational number, but does not contain all the rationals,

(ii) if $r \in \alpha$, then every rational number less than r is also in α.

(iii) α contains no greatest member.

A set of rational numbers which satisfies the above properties divides the set Q of all rationals into two classes: the set α which is called the *lower class* and the set of rational numbers not in α, called the *upper class*.

Note that the lower class, by definition, has no greatest member. But the upper class may or may not have a smallest element. If the upper class has a smallest member, the section is called a *rational section*. If $r \in Q$, and if we define the set α by

$$\alpha = \{x \in Q : x < r\}$$

then α is a rational section. Observe that if $x \in \alpha$, then $\frac{1}{2}(x + r) \in \alpha$ since $x < \frac{1}{2}(x + r) < r$. Now α is a rational section and $r \in \alpha$. So r is the least member of the upper class corresponding to α.

Each section of the rational numbers defines a real number. A rational section correspond to or is identified with a rational number. All other sections define irrational numbers. In the sense of correspondence mentioned above the set of all rational numbers forms a subset of the set of all real numbers.

We now introduce some definitions.

1.3.1. Definitions

(a) Two sections α and β are said to be equal if they are equal as sets (of rationals).

(b) If $\alpha \neq \beta$ then either $\alpha < \beta$ or $\alpha > \beta$. $\alpha < \beta$ if there is some $b \notin \beta$ such that $b \in \alpha$. $\alpha > \beta$ is defined similarly.

(c) The *sum* of two sections α and β is the set γ defined by

$$\gamma = \{c \in Q : c = a + b \,,\, a \in \alpha \,,\, b \in \beta\}$$

and we write this as $\gamma = \alpha + \beta$. We need to show that γ is, in fact, a section.

Since α and β are sections, there exist a rational $p \in \alpha$ and a rational $q \in \beta$. Then for any $a \in \alpha$ and $b \in \beta$, $a + b < p + q$, showing that $p + q \in \gamma$.

Now let $a \in \alpha$, $b \in \beta$, $c = a + b \in \gamma$. For any rational number $p < c$ we can find a rational number q such that $p = q + b$. Hence $q < a$

and $q \in \alpha$ and so $p \in \gamma$. Let $c = a + b$ for some $a \in \alpha$ and $b \in \beta$. Since α is a section there is a rational number p such that $p > \alpha$ and $p \in \alpha$. Then $p + b \in \gamma$ and $p + b > c$. This means that γ has no greatest member.

This shows that γ is a section.

Observe that addition of sections is associative and commutative and this follows from the fact that addition of rational numbers is associative and commutative.

(d) The product of two sections α and β is the set γ defined by

$$\gamma = \{c \in Q : c = ab, a \in \alpha, b \in \beta\}$$

By using arguments similar to addition it can be shown that γ is a section.

Further, since multiplication of rationals is associative and commutative, multiplication of cuts also satisfies these properties.

Also distributive laws for addition and multiplication of cuts follow from the corresponding properties of rational numbers.

1.3.2. Theorem (Dedekind)

Let the set R of all real numbers be divided into two classes L and U such that

(a) $L \cup U = R$

(b) $L \cap U = \phi$,

(c) $L \neq \phi$, $U \neq \phi$,

(d) every element of L is less than every element of U.

Then there exists a unique (one and only one) real number γ, such that every real number less than γ belongs to L and every real number greater than γ belongs to U.

The number γ may belong to either L or U. If it belongs to L it is the largest member of L; if it belongs to U, it is the smallest member of U.

Let γ be the set of all rationals a such that a is in α for some α in L. Then:

(a) Since $L \neq \phi$, $\gamma \neq \phi$, γ does not contain all the rationals. Let $\beta \in U$ and p a rational not in β, then p cannot be in any α in L, for $\alpha < \beta$ and $p > a$ for every rational a in α.

(b) Let a be in γ, then a is in α for some $\alpha \in L$. Then if $p < a$, p is in α and so p is in γ.

(c) Let a be in γ, then a is in α for some $\alpha \in L$. Since α is a section, α has no greatest member, Hence γ has no greatest member.

Thus γ is a section of rationals, in other words, γ is a real number. Also it is unique; for if there were two such numbers $\gamma_1, \gamma_2, \gamma_1 < \gamma_2$, we could find a number γ_3 such that $\gamma_1 < \gamma_3 < \gamma_2$. Now since $\gamma_1 < \gamma_3$, it follows that $\gamma_3 \in U$. Again, from $\gamma_3 < \gamma_2$ we obtain $\gamma_3 \in L$ and this contradicts (b).

We represent the rational numbers geometrically by points on a straight line by choosing any point on the line as the number 0, which is called the origin, and a suitable length from the origin as the unit length, In this manner all the rational numbers can be plotted on a line. After all rationals are plotted, there will still be gaps on the line, in the sense that there will be many points on the line which are not represented by rational numbers. *Dedekind's theorem* says that those points on the line which do not represent rationals represent irrational numbers. Every point on a line represent a real number and conversely every real number corresponds to a point on a line. Thus unlike in the case of rationals there are no gaps in the set of all real numbers; in other words, the set of all real numbers is complete. This is called the *completeness property* of the real numbers. Several other consequences and equivalences of this important property of the real numbers will be discussed in the sequel.

1.4 FIELD AND ORDER AXIOMS

The two operations, addition and multiplication in R, are mappings from $R \times R$ into R and hence are called binary operations. The following properties hold in the set of all real numbers : For all $a, b, c,$ ε R,

1. $a + b \, \varepsilon \, R$ (the closure law)
2. $a + b = b + a$ (commutative law)
3. $a + (b + c) = (a + b) + c$ (associative law)
4. There exists an element $0 \, \varepsilon \, R$ such that $a + 0 = a$ (existence of additive identity)
5. For every $a \varepsilon R$, there exists an element $(-a) \in R$ such that $a + (-a) = 0$.
6. $ab \in R$ (The closure law for multiplication)
7. $ab = ba$ (commutative law)

8. $a(bc) = (ab)c$ (associative law)
9. There exists an element $1 \varepsilon R$, $1 \neq 0$ such that $a1 = a$ (multiplicative identity)
10. For every $a \varepsilon R$, $a \neq 0$ there exists **an** element $a^{-1} \varepsilon R$ such that

$$aa^{-1} = 1. \text{ (multiplicative inverse of nonzero elements)}$$

Properties 1 and 6 are superfluous statements. These are just the statements that addition and multiplication are maps from $R \times R$ into R. In other words R is a field under addition and multiplication of real numbers.

The difference of a and b is defined by $a + (-b)$ and is written as $a - b$. Similarly the quotient of a and b, $b \neq 0$ is defined by ab^{-1} and is written as $\frac{a}{b}$.

1.4.1. Definition

A field F is said to be an ordered field if there exists a nonempty subset P of F such that

(i) $P \cup \{0\} \cup (-P) = F$

where $(-P) = \{(-x) : x \varepsilon P\}$

In other words, for any a in an ordered field F one and only one of the following statements is true:

(i) $a \varepsilon P$, $-a \varepsilon P$, $a = 0$

(ii) If a, $b \varepsilon P$ then $a + b \varepsilon P$

(iii) If a, $b \varepsilon P$ then $ab \varepsilon P$

If F is an order field we define a binary relation in it as follows : For a, $b \varepsilon F$, $a > b$ iff $a - b \varepsilon P$. In particular, $a > 0$ iff $a \varepsilon P$

Therefore the set P is called the set of all positive elements in F.

We introduce the following relations:

$a \geq b$ iff $a > b$ or $a = b$.

and $a \leq b$ iff $a < b$ or $a = b$.

1.4.2. Theorem

The relation $\leq$ defined above is a totally ordered relation in F. In other words,

(i) $a \leq a$ (reflexive)

(ii) If $a \leq b$, $b \leq a$ then $a = b$ (antisymmetry)

(iii) If $a \leq b$, $b \leq c$ then $a \leq c$ (transitivity)

(iv) For $a, b \in F$, one and only one of the following statements hold: $a < b$, $a = b$, $b < a$.

The easy proof is left to the reader as an exercise.

1.4.3. Axiom of Order

The field of all real numbers R is an ordered field with $P = \{x \in R : x > 0\}$.

This is known as the axiom of order.

1.4.4. Absolute Value

Let $x \in R$, We define the absolute value of x, written as $|x|$, *by*

$$|x| = \begin{cases} x \text{ if } x \geq 0 \\ -x \text{ if } x < 0 \end{cases}$$

We have

1.4.5. Theorem

For $a, b, \in R$

(i) $|a| \geq a$, $|a| \geq -a$

(ii) $|a + b| \leq |a| + |b|$

Proof: (i) Suppose $a \geq 0$. By definition, $|a| = a$.

Moreover, $-a \leq 0$, so $|a| \geq 0 \geq -a$.

Suppose $a < 0$. Then $|a| = -a$ and $|a| \geq 0 > a$.

Thus (i) holds for all real numbers.

(ii) If $|a| \geq a$ and $|b| \geq b$, then $|a| + |b| \geq a + b$.

If $|a| \geq -a$ and $|b| \geq -b$, then $|a| + |b| \geq -a - b = -(a + b)$.

By definition $|a + b|$ is equal to either $a + b$ or $-(a + b)$. *In either case*, $|a| + |b| \geq |a + b|$ and this completes the proof.

1.4.6. Definition

Let S be a nonempty subsets of R. An element $b \in R$ is said to be an upper bound for S if $x \leq b$ for every $x \in S$.

1.4.7. Example

(a) The set $S = \{1, 2, 3, 6, 7, 9\}$ has 10 as an upper bound. Any number greater than or equal to 9 is an upper bound.

(b) $S = \{x \varepsilon R : x$, is positive$\}$ does not have an upper bound.

(c) $S = \left\{x : x = \frac{n}{n+1}, n = 1, 2, \ldots.\right\}$ has 1 as an upper bound.

1.4.8. Definition

A set $S \subset R$ is said to be *bounded above* if it has an upper bound.

A number $c \varepsilon R$ is said to be the *least upper bound* (lub, in short) or *supremum* (sup, in short) of a set S if (i) c is an upper bound, (ii) if b is an upper bound other than c, then $b > c$.

1.4.9. Remark

(a) 9 is the lub of Example (1. 4.7) (a) and 1 is the lub of Example (1.4.7) (c).
(b) lub of a set which is bounded above is unique.

1.4.10. Axiom of order completeness of *R*

Any nonempty set of real numbers which is bounded above has a lub in *R*.

Lower bounds and greatest lower bounds (glb) in short) or infimum (inf in short) of nonempty subsets of *R* are defined in a similar manner. For example, $b \varepsilon R$ is called lower bound for a nonempty set $S \subset R$ if $x \geq b$ for all $x \varepsilon S$. $c \varepsilon R$ is called the glb of S if (i) c is a lower bound for S, (ii) for any lower bound b other than c, $b < c$.

The axiom of order completeness can also be formulated in the following equivalent form.

1.4.11. Any nonempty set of real numbers which is bounded below has glb in *R*

Either one or both the above axioms are called the *axioms of order completeness.* There are also other equivalent forms of completeness, some of which do not involve order structure. One of them is Canchy completeness (i.e. every Cauchy sequence of real numbers is convergent) and this will be discussed in Chapter 3. We shall not make any attempt to prove the equivalence of various forms of the completeness axiom in this introductory book.

1.4.12. Definition

An ordered field F is said to be *archimedean ordered* if for $b, c \in F, c > 0$, there exists a positive integer n such that

$$nc = \underbrace{c + c + \ldots + c}_{n \text{ times}} > b.$$

1.4.13. Theorem

The ordered field R is archimedean ordered, i.e., if $a, b, \in R$, then there exists a positive integer n such that $na > b$.

Proof: Let $a, b, \in R$. Assume, to the contrary, that for all positive integers n, $na < b$. Thus the set $S = \{na : n \text{ is a positive integer}\}$ is bounded above, b being an upper bound. By the completeness property of R, S must have supremum, say M. Therefore, for all $n \in N$,

$$na \leq M$$

$$\Rightarrow (n+1)\, a \leq M$$

$$\Rightarrow na \leq M - a,$$

i.e. $M - a$ is an upper bound of S. Thus a number less than the sup is an upper bound and this contradiction proves the result.

The real number system can thus be precisely described as complete archimedean ordered field. But any two complete archimedean ordered fields are indistinguishable in the sense that there exists a 1–1 (one to one) correspondence between them which also preserves the binary operations. We assume this fact without proof and define the real number system to be the complete archimedean ordered field.

1.4.14. Decimal Representation

Let x be any positive real number. Then by Theorem 1.4.13 we can find an integer n such that

$$n \leq x < (n+1)$$

Let d_i $(i \in N)$ be the digits from the set

$$0, 1, 2, 3, 4, 5, 6, 7, 8, 9.$$

Consider the set E where

$$E = \left\{ n + \frac{d_1}{10} + \frac{d_2}{10^2} + \ldots + \frac{d_k}{10^k} : k = 0, 1, 2, \ldots\ldots \right\}$$

Then

$$x = \text{lub } E$$

Thus we can write

$$x = n + \frac{d_1}{10} + \frac{d_2}{10^2} + \ldots.$$

$$= n \,.\, d_1\, d_2\, d_3 \ldots.$$

This is called the decimal representation of a real number.

It may be seen that all the rational numbers will have a finite number of infinite recurring digits after the decimal point in their decimal representation.

1.5 COUNTABLE AND UNCOUNTABLE SETS

We know from our experience that before a child formally, learns counting, he intuitively masters the art of matching or putting two sets of objects in 1–1 *correspondence* or what is also called a *bijective correspondence*. We shall now formally describe concept of matching in the form of a definition.

1.5.1. Definition

Let A and B be any two sets. A and B are said to be equivalent and written as $A \sim B$, if there exists a bijection

$$f : A \to B$$

If A is a finite set with n elements and $A \sim B$, then B has exactly n elements also. It is also true that if A is a finite set and $B \subset A$, then B cannot be equivalent to A. But this fails to be true for infinite sets. We shall now illustrate this by two simple examples.

1.5.2. Examples

(a) Let A denote the set of all even positive integers. Then $A \subset N$. But

$$f : N \to A$$

defined by $f(n) = 2n$

is a bijection and hence $N \sim A$.

(b) $N \subset Z$

But $f : Z \to N$ defined by

$$f(n) = \begin{cases} 0 & n = 0 \\ -2n - 1 & n < 0 \\ 2n & n > 0 \end{cases}$$

is a bijection and thus

$N \sim Z.$

This shows that we have to classify infinite sets in the following manner.

1.5.3. Definition

A infinite set is *countably infinite* if it is equivalent to N. A set is countable if either it is finite or countably infinite.

A set which is not countable is called *uncountable.*

1.5.4. Theorem

Every subset of a countable set is countable.

Proof: Let A be countable and $B \subset A$. We have to show that B is countable. If either A or B is finite, then clearly B is countable. Now assume that A is infinite and B is an infinite subset of A. Since A is countable we can write

$$A = \{x_1, x_2, x_3, \ldots, x_n, \ldots\}$$

Let the mapping $f: N \to B$ be defined by $f(1) = x_{n_1}$, where n_1 is the smallest positive integer such that $x_{n_1} \in B$. Let

$$f(2) = x_{n_2}$$

where n_2 is the smallest positive integer greater than n_1 such that $x_{n_2} \in B$. We continue in this way and in general write

$$f(k) = x_{n_k}$$

where n_k is the smallest positive integer greater that $x_{n_{k-1}}$ such that $x_{n_k} \in B$. Thus there exists a 1–1 correspondence between N and B hence B is countable.

1.5.5. Theorem

The countable union of countable sets is countable.

Proof: Let $\{A_1, A_2, \ldots, A_n, \ldots\}$ be a countable family of countable sets. Let

$$A = \bigcup_{n=1}^{\infty} A_n.$$

We have to show that A is countable. Since each A_n is countable we can write

$$A_1 = \{x_{11}, x_{12}, x_{13}, \ldots\}$$

$$A_2 = \{x_{21}, x_{22}, x_{23}, \ldots\}$$

$$A_3 = \{x_{31}, x_{32}, x_{33}, \ldots\}$$

The set A precisely consists of all these elements listed above and we can arrange these elements as indicated by rows in the following manner:

1.5.6.

$x_{11}; x_{21}, x_{12}; x_{31}, x_{22}, x_{13}; \ldots$

If no two of the sets A_n have elements in common, then the elements in the above list are all distinct and there exists a 1–1 correspondence between $A = \bigcup_{n=1}^{\infty} A_n$ and the set n of positive integers which is given as follows where the images are noted under the objects:

$$\begin{matrix} x_{11} & x_{21} & x_{12} & x_{31} & x_{22} & x_{13} & \cdots \\ 1 & 2 & 3 & 4 & 5 & 6 & \ldots \end{matrix}$$

Hence A would be countable. If any two of the sets A_n have common elements, these will appear more than once in the list (1.5.6). Hence there exists a subset B of N such that A and B are equivalent. Now, since B is a subset of the countable set N, it is countable and hence A is countable.

1.5.7. Theorem

The set R of all real numbers is uncountable.

Proof: First we consider that open interval (0, 1) which is a subset of R. Using decimal representation of real numbers and assuming that the set of real numbers in (0, 1) are countable, we write

$(0, 1) = A$ and

$$A = \{x_n : 0 < x_n < 1 \,;\, n = 1, 2, 3, \ldots\}$$

where

$$x_1 = 0 \cdot d_{11}\, d_{12}\, d_{13} \ldots$$

$$x_2 = 0 \cdot d_{21}\, d_{22}\, d_{23}\ \ldots$$

$$\vdots$$

$$x_n = 0 \cdot d_{n1}\, d_{n2}\, d_{n3}\ \ldots$$

where $d_{ij} \in \{0, 1, 2, 3, 4, 5, 6, 7, 8, 9, \} \equiv D$

for all $i = 1, 2, 3, \ldots..$ and $j = 1, 2, 3, \ldots...$ Now we construct a real number x_0 where

$$x_0 = 0 \cdot d_1\, d_2\, d_3 \ldots$$

$$d_i \in D,\ i = 1, 2, 3, \ldots$$

Here $d_i = 1$ if $d_{ii} \neq 1$ and $d_i = 2$ if $d_{ii} = 1$ for all $i = 1, 2, 3, \ldots$ Thus $x_0 \in (0, 1)$. But

$$x_0 \neq x_i \text{ for all } i = 1, 2, 3, \ldots$$

implying $x_0 \notin A$. this contradiction shows that (0, 1) is uncountable and R, being a superset of (0, 1), is also uncountable by 1.5.4.

1.5.8. Corollary

Any interval (a, b) on R is uncountable. This can be proved with the help of the result that (0, 1) is uncountable and the transformation

$$f(x) = a + (b - a)\, x$$

which maps (0, 1) onto (a, b) (having one–one correspondence).

1.6 SETS AND INTERVALS

The set A defined by

$$A = \{x \in R : a < x < b\}$$

where a and b are reals, is called an *open interval* and is denoted by (a, b). Similarly the set

$$A = \{x \in R : a \leq x \leq b\}$$

is a *closed interval* which is denoted by $[a, b]$.

We have *semi-closed* or *semi-open* intervals also. The interval $(a, b]$ is 'open on the left and closed on the right', while the interval $[a, b)$ is 'closed on left and open on right'. These intervals denote the sets $\{x : a < x \leq b\}$ and $\{x : a \leq x < b\}$ respectively.

1.6.1. Definition

A set N of real numbers is called δ *Neighbourhood* of a point a if there is an open interval $(a-\delta, a+\delta)$, where $\delta > 0$,
such that

$$(a-\delta, a+\delta) \subset N$$

An open interval is a neighbourhood of each of its points. If N_1 and N_2 are neighbourhoods of a point a then $N_1 \cup N_2$ and $N_1 \cap N_2$ are also neighbourhoods of a. A point x of a set is said to be its *Interior point* if there exists a δ such that

$$(x-\delta, x+\delta) \subset E$$

A set E of real numbers is said to be open if each of its points are interior points. An open interval is an open set.

Let E be a subset of R and $x \in R$. If each neighbourghood of X contains at least one point of E other than x, then x is said to be a *limit point* of E ; x may or may not belong to E.

If all the limit points of E belong to it then E is said to be a *closed set.*

The complement of a closed set is an open set.

A point which is not a limit point of a set is an *isolated* point of that set.

1.6.2. Theorem (Bolzano-Weierstrass)

Every infinite bounded set has a limit point.

Proof: Let E be any infinite bounded set of real numbers and α and β its greatest lower bound and least upper bound respectively. Then we define a set A as follows:

A point x of E belongs to A if and only if it exceeds at the most a finite number of points of E. This implies that A is nonempty as $\gamma \in A$. Also as β is the upper bound of A, no number greater than β can belong to A. Now $A \subset R$ and R has the least upper bound property. Therefore A must have least upper bound, say μ. We consider a neighbourhood $N = (\mu - \in, \mu + \in)$, $\in > 0$, of μ.

As μ is the least upper bound of A, there will be at least one point, say y such that $y \in A$ and $y > \mu - \in$. Also it should exceed at most a finite number of points of E and consequently the point $\mu - \in$ can exceed at most a finite number of points of E.

Also the point $\mu + \epsilon$ cannot belong to A and consequently, $\mu + \epsilon$ must exceed an infinite number of points of E. This means the neighbourhood N contains infinitely many points of E. Hence μ is a limit point of E. This proves the theorem.

More discussion on point set theory will be presented in the following chapters.

1.7 COMPLEX NUMBERS

A complex number z is an ordered pair (a, b) of real numbers. The real numbers a and b are respectively called the *real* and *imaginary* parts of z. Symbolically we write $a = \text{Re } z$ and $b = \text{Im } z$. Two complex numbers $z_1 = (a_1, b_1)$ and $z_2 = (a_2, b_2)$ are said to be equal if and only if $a_1 = a_2$ and $b_1 = b_2$. We write C for the set of all complex numbers. Addition and multiplication in C are defined as follows:

$$z_1 + z_2 = (a_1 + a_2, b_1 + b_2),$$

$$z_1 + z_2 = (a_1a_2 - b_1b_2 , a_1b_2 + a_2b_1).$$

With addition and multiplication defined as above, C is a field. It is easy to see that addition and multiplication satisfy the associative, commutative and distributive laws and we leave the checking of these laws as an exercise for the reader. The complex numbers (0, 0) and (1, 0) are the additive and multiplicative identities in C. The operations of subtraction and division in C are defined as

$$z_1 - z_2 = (a_1 - a_2, b_2 - b_2),$$

$$\frac{z_1}{z_2} = \left(\frac{a_1 a_2 - b_1 b_2}{a_2^2 + b_2^2}, \frac{a_2 b_1 - a_1 b_2}{a_2^2 + b_2^2}\right) \text{ if}$$

$$z_2 = (a_2, b_2) \neq (0, 0).$$

Note that for $a, b \in R$, the following properties hold:

$$(a, 0) + (b, 0) = (a + b, 0),$$

$$(a, 0)\ (b, 0) = (ab, 0),$$

$$\frac{(a, 0)}{(b, 0)} = \left(\frac{a}{b}, 0\right) \text{ if } b \neq 0$$

We can identify the complex number (a, o) with the real number a and by this identification R may be considered as a subset of C. Observe that the complex number (a, b) can be written in a more convenient way as $a + ib$ where $i = (0, 1)$. For, when we identify $(a, 0)$ and $(b, 0)$ with a and b respectively, then

$$a + ib = (a, 0) + (0, 1)(b, 0) = (a, 0) + (0, b)$$
$$= (a, b)$$

Note that $i^2 = (0, 1)(0, 1) = (-1, 0) = -1$. It should be noted that the equation $z^2 + 1 = 0$ has no solution in R, but this is solvable in C, the roots being $\pm i$. In fact any equation with complex coefficients has a root (in fact all the roots) in C. The proof of this uses ideas of complex analysis and hence we shall not prove it here.

1.7.1. Definition

If $z = (a, b)$, then the unique nonnegative real number $\sqrt{a^2 + b^2}$ is called the *absolute value* or *modulus* of z and is written as $|z|$. The complex number $(a, -b)$ is called the *conjugate* if z and is written as $\bar{z}$.

1.7.2. Theorem

If $z, z_1, z_2, \varepsilon C$ then

(a) $|z| \geq 0$, $|z| = 0$ iff $z = 0$

(b) $|z_1 z_2| = |z_1|\ |z_2|$

(c) $|z_1 + z_2| \leq |z_1| + |z_2|$

(d) $\overline{z_1 + z_2} = \overline{z_1} + \overline{z_2}$

(e) $\overline{z_1\ z_2} = \overline{z_1}\ \overline{z_2}$

(f) $\overline{\left(\dfrac{z_1}{z_2}\right)} = \dfrac{\bar{z}_1}{\bar{z}_2}$, $|z_2| > 0$

(g) $|\bar{z}| = |z|$

(h) $z\bar{z} = |z|^2$

(i) $z + \bar{z}$ is real

(j) $z\ \varepsilon R \Rightarrow \bar{z} = z$

(k) $2Re z = z + \bar{z}$, $2i \operatorname{Im} z = z - \bar{z}$

(l) $\bar{\bar{z}} = z$

We only prove (c). For $z_1, z_2, \varepsilon C$,

$$|z_1 + z_2|^2 = (z_1 + z_2)\,(\overline{z_1 + z_2}) = (z_1 + z_2)\ (\bar{z}_1 + \bar{z}_2)$$
$$= z_1\,\bar{z}_1 + z_2\,\bar{z}_2 + z_1\,\bar{z}_2 + \bar{z}_1\,z_2$$

$$= |z_1|^2 + |z_2|^2 + 2\ \text{Re}\ (z_1\bar{z}_2)$$

$$\leq |z_1|^2 + |z_2|^2 + 2\ |z_1\bar{z}_2|$$

$$= |z_1|^2 + |z_2|^2 + 2\ |z_1|\ |z_2|$$

$$= (|z_1| + |z_2|)^2$$

1.7.3. Theorem

The field C is not ordered.

Proof: Assume, to the contrary, that there exists a subset P of C with $i \varepsilon P$ or, $-i \varepsilon P$.

If $i \varepsilon P$, then $i^2 = -1\ \varepsilon\ P$. This is a contradiction.

If $-i\ \varepsilon\ P$,Then

$(-i)^2 = -1\ \varepsilon\ P$. This is also a contradiction.

EXERCISES

1. Answer 'Yes' or 'No':

(i) The equation $x^2 = n$ is not satisfied by a rational number.

(ii) The set Z of all the integers satisfies all the field axioms.

(iii) The set Q of rational numbers is divided by a Dedekind cut.

(iv) The set of rational numbers has no gaps.

(v) The set of real number has gaps.

2. State whether true or false:

(i) Any bounded subset of Q always has least upper bound and greatest lower bound in Q.

(ii) The set I of irrational numbers satisfies the completeness property (also known as least upper bound property).

(iii) $I \cap Q = \phi$

(iv) $Z \cup R = Z$

(v) $R \cup (I \cap Q) = I$

3. State which of the following are correct:

(i) $a > b \Rightarrow \frac{1}{a} < \frac{1}{b}$

(ii) $a < b \Rightarrow a + c < b + c$ for all $a, b, c, \in R$

(iii) $a < b \Rightarrow a \leq b$

(iv) An irrational number has no decimal representation.

(v) $A = \{x : 0 < x < 1\}$ is countable.

(vi) $A = z \cup \left\{\frac{1}{n} : n \in N\right\}$ where N is set of all positive integers, is uncountable.

(vii) $A = \bigcup_{i=1}^{\infty} \left\{x : 1 - \frac{1}{i} < x < \frac{1}{i}\right\}$ is uncountable.

(viii) The set I of all irrational numbers is countable.

(ix) $A = \{x : a < x < b;\ a \text{ and } b \text{ are rationals}\}$ has no limit point.

(x) The set$\{-1, 0, 1\}$ is closed.

(xi) The set C of all complex numbers is an ordered field.

(xii) $z \cdot \bar{z}$ is real where $z \in C$.

(xiii) $R \subset Z$

4. If z_1 and z_2 are complex numbers prove that

$|z_1| - |z_2| \le |z_1 - z_2|$

5. If $z_1, z_2, \ldots, z_n$ are complex numbers prove that

$|z_1 + z_2 + \ldots + z_n| \le |z_1| + |z_2| + \ldots + |z_n|$

6. Give examples of sets which are:

(i) bounded

(ii) not bounded

(iii) bounded below but not bounded above

(iv) bounded above but not bounded below

7. Find the infimum and supremum of the following sets;

(i) $\left\{\frac{1}{n} : n \in N\right\}$

(ii) $\left\{\frac{(-1)^n}{n} : n \in N\right\}$

(iii) $\left\{1 + \frac{(-1)^n}{n} : n \in N\right\}$

8. Let A be a nonempty set of real numbers which is bounded below.

Let

$$-A = \{-x : x \in A\}$$

Show that

$$\text{Inf } A = -\sup(-A).$$

9. Prove that $|x| = \max(x, -x)$, $x \in R$

10. Find the greatest lower bound and least upper bound of A, where

$$A = \left\{\frac{n+3}{n+1} : n \in N\right\}$$

11. Prove that the set A where

$$A = \{n \cdot d_1 d_2 :$$

where $n \in N$ and d_1 and d_2 are any single digits} is countable.

12. For $z = (a, b)$, a and b are nonzero reals, find z^{-1} such that

$$z z^{-1} = 1$$

13. For $x \in R$, show that

$$|x| < a \Rightarrow -a < x < a.$$

14. Prove that an infinite subset of a countable set is countable.

15. In the set C of complex numbers define a relation '$\leq$' as follows $z_1 = (x_1, y_1,) \leq z_2 = (x_2, y_2,)$ iff $x_1 \leq x_2$ and $y_1 \leq y_2$. Show that this is a partially ordered relation but is not a linearly order relation on C.

16. If A and B are bounded subsets of R, then show that $A \cap B$ and $A \cup B$ are bounded. Moreover

$$\sup(A \cup B) = \max(\text{sub } A, \sup B)$$

$$\inf(A \cup B) = \min(\inf A, \text{Inf } B).$$

17. Show that there exists no rational number whose square is 3.

18. Show that the glb (greatest lower bound) of a subset of R which is bounded below is unique.

19. Show that [0, 1] is uncountable.

20. Show that (0, 1) and R are equivalent.

[*Hint*: Define a map $g : (0, 1) \to R$ by $g(x) = \tan \frac{\pi}{2}(2x - 1)$ and show that g is 1–1 and onto]

21. If A is a non-empty bounded subset of R, and B is the set of all upper bounds for A, prove that

$$\text{glb } B = \text{lub } A$$

22. If $f : A \to B$ and the range of f is uncountable, prove that the domain of f is uncountable.

2 Metric Spaces

1.1. DEFINITION AND EXAMPLES

2.1.1. Definition

Let X be any nonempty set. A metric on X is a mapping d $: X \times X \to R$ (where R denotes the set of all real numbers) which satisfies the following rules:

For $x, y, z \in X$,

(i) $d(x, y) \geq 0$, $d(x, y) = 0$ iff $x = y$

(ii) $d(x, y) = d(y, x)$,

(iii) $d(x, z) \leq d(x, y) + d(y, z)$

A set X together with a metric d defined on it is called a metric space and it is denoted by (X, d). We sometimes write X to denote a metric space if there is no chance of confusion.

We now consider some examples of metric spaces.

2.1.2. Examples

(a) Let X be a nonempty set. Then it can be made into a metric space in a trivial way. Define d $: X \times X \to R$ by

$$d(x,y) = \begin{cases} 0 \text{ if } x = y \\ 1 \text{ if } x \neq y \end{cases}$$

It is left to the reader to verify that d defines a metric on X and it is called the *trivial metric*.

(b) The real line R is a metric space with the metric defined by

$$d(x, y) = |x - y|$$

for $x, y \in R$ where $|x - y|$ denotes the absolute value of $x - y$. In fact $|x - y|$ denotes the distance between the points x and y on the real line and this metric is called the *usual metric* or *standard metric* on R. The reader will become familiar in due course with the fact that other metrics can also be defined on R, that in fact, the trivial metric is also a metric on R.

(c) The two-dimesional plane R^2 and the set of all complex numbers C are indistinguishable in the sense that there exists a 1–1 correspondence between them, the map which associates any point $x = (x_1, x_2) \varepsilon R^2$ with $x = x_1 + i x_2 \varepsilon C$ is a bijective mapping. Therefore we identify R^2 with C and call it the *complex plane*. It is left to the reader as an exercise to verify that C is a metric space with the metric d defined by

$$d(x, y) = |x - y|$$

for $x, y \varepsilon C$. Recall that for $x = (x_1, x_2)$ and $y = (y_1, y_2)$

$$d(x,y) = |x - y| = [(x_1 - y_1)^2 + (x_2 - y_2)]^{1/2}$$

This metric is called the *usual metric* or *standard metric* on C or on R^2. In this metric, the set

$$\{x \in R^2 : d(x, 0) < r\}$$

denotes the set of all points within the circle of radius r and centre at the origin (Fig. 2.1),

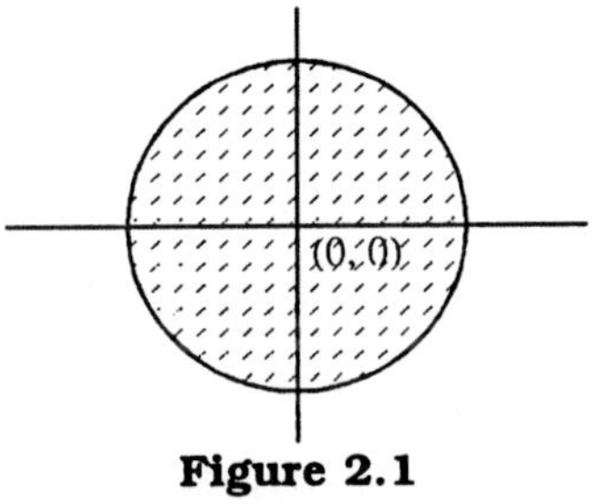

Figure 2.1

We shall now give some further examples of metrics on R^2 which are different from the usual distance between two points. These examples will also convince readers that there can be defined more than one metric on the same set giving rise to different metric spaces.

2.1.3. Examples

(a) Let $x = R^2$. For $x = (x_1, x_2)$ and $y = (y_1, y_2)$ $\in R^2$ define

$$d(x,y) = \max(|x_1 - y_1|, |x_2 - y_2|),$$

where 'max' denotes the maximum. For $x, y, z \in R^2$ observe that for $i = 1, 2$

$$|x_i - z_i| = |(x_i - y_i) + (y_i - z_i)| \leq |x_i - y_i| + |y_i - z_i|$$

Therefore

$$\max(|x_1 - z_1|, |x_2 - z_2|) \leq \max(|x_1 - y_1|, |x_2 - y_2|) + \max(|y_1 - z_1|, |y_2 - z_2|)$$

and so

$$d(x, z) \leq d(x, y) + d(y, z).$$

The reader should verify rules (i) and (ii) and convince himself that d is a metric on R^2. this is called the *square metric on* R^2.

For any $x \in R^2$ let us now represent graphically the set

$$S_r = \{x \in R^2 : d(x,0) < r\}$$

where 0 denotes $(0, 0) \in R^2$ and $r > 0$.

Observe that if $x \in S_r$, then

$$\max(|x_1|, |x_2|) < r.$$

So that $|x_1| < r$ and $|x_2| < r$. That is,

$$-r < x_1 < r, \ -r < x_2 < r.$$

Therefore x must be a point inside the square whose vertices are given by $(-r, -r)$ $(r, -r)$, (r, r) and $(-r, r)$ and hence the name square metric is justified (Fig. 2.2).

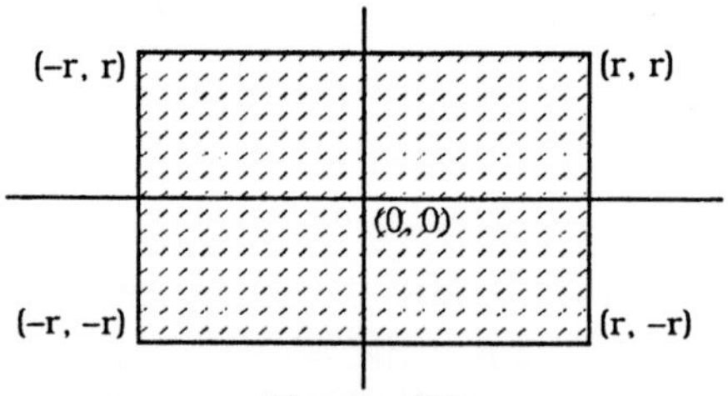

Figure 2.2

(b) Let $X = R^2$ and for $x, y \in R^2$ define

$$d(x,y) = |x_1 - y_1| + |x_2 - y_2|.$$

This also defines a metric on R^2. We shall prove only the triangle inequality, leaving the verification of other rules as an exercise for the reader. Let $x, y, z \in R^2$. Then

$$\begin{aligned} d(x, z) &= |x_1 - z_1| + |x_2 - z_2| \\ &= |x_1 - y_1 + y_1 - z_1| + |x_2 - y_2 + y_2 - z_2| \\ &\leq (|x_1 - y_1| + |y_1 - z_1| + (|x_2 - y_2| + |y_2 - z_2|) \\ &= (|x_1 - y_1| + |x_2 - y_2|) + (|y_1 - z_1| + |y_2 - z_2|) \\ &= d(x, y) + d(y, z). \end{aligned}$$

In this metric the set

$$S_r = \{x \varepsilon R^2 : d(x, 0) < r\}$$

is the set of all points inside the square whose vertices are

$(r, 0)$, $(0, r)$ $(-r, 0)$ and $(0, -r)$ (Fig. 2.3)

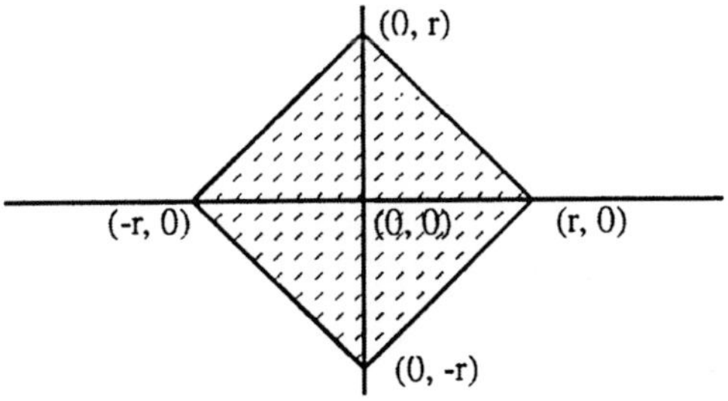

Figure 2.3

The set S_r in R^2 corresponding to each of the metrics defined in Examples 2.1.2 (c), 2.1.3 (a) and 2.1.3 (b) can be graphically represented in a single diagram as shown in Fig. 2.4.

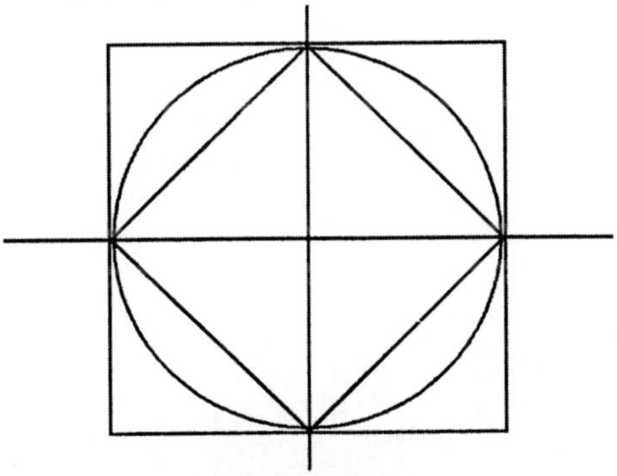

Figure 2.4

Throughout the rest of this book, unless otherwise stated. R will denote the 'real' line equipped with its standard metric. Similarly C or R^2 will stand for the complex plane with its standard metric.

2.2 OPEN SETS

2.2.1. Definition

Let (X, d) be a metric space. Let $a \varepsilon X$ and $r > 0$. The set

$$S_r(a) = \{x \varepsilon X : d(x, a) < r\}$$

is called a *neighbourhood* (or an *open sphere* or an *open ball*). The point a is called the centre and r is called the radius of $S_r(a)$

Note that $S_r(a)$ is always nonempty, since $d(a, a) = 0 < r$ so $a \varepsilon S_r(a)$.

We now give some examples of open balls.

2.2.2. Examples

(a) Let (X, d) be the real line R (with the standard metric). Let $a \varepsilon R$ and $r > 0$. Then

$$\begin{aligned} S_r(a) &= \{x \varepsilon R : |x - a| < r\} \\ &= \{x \varepsilon R : -r < x - a < r\} \\ &= \{x \varepsilon R : a - r < x < a + r\} \\ &= (a - r, a + r) \end{aligned}$$

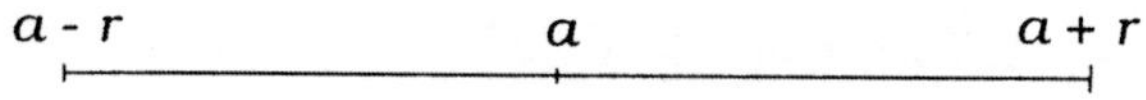

Figure 2.5

(b) Let $x = R^2$ with the standard metric. Let $a = (a_1, a_2) \varepsilon R^2$ and $r > 0$. Then

$$S_r(a) = \{x = (x_1, x_2) \varepsilon R^2 : (x_1 - a_1)^2 + (x_2 - a_2)^2 < r^2\}$$

and this is the set of all points inside the circle of radius r and centre a (Fig. 2.6).

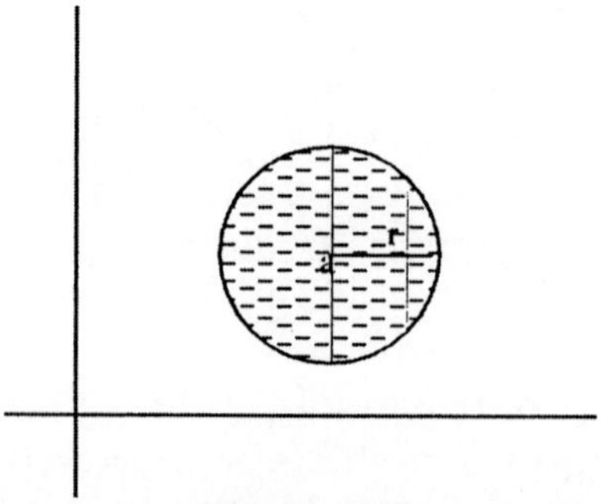

Figure 2.6

(c) let X be any set with trivial metric d. Let $a \varepsilon X$. In this case $d(x,a) = 1$ if $x \neq a$. Therefore it follows that $S_r(a) = \{a\}$ if $0 < r \leq 1$ and $S_r(a) = X$ if $r > 1$.

2.2.3 Definition

Let (X, d) be a metric space and let $A \subset X$. A point $a \varepsilon A$ is said to be an interior point of A if there exists an $r > 0$ such that $S_r(a) \subset A$. A set $G \subset X$ is said to be an open set if every point of G is an interior point of G. In other words, G is open if for every $a \varepsilon G$ there exists an *$r > 0$ such that $S_r(a) \subset G$.*

2.2.4. Examples

(a) Let $X = R$. Let $A = (0, 1) \subset R$. Then every point of A is an interior point of A and hence A is an open set. To prove this, let $a \varepsilon A$. Since $a \neq 0$, $a \neq 1$,

$$|a - 0| = r_1 > 0, \ |a - 1| = r_2 > 0$$

Choose $r = \min(r_1, r_2) > 0$. Then clearly

$$S_r(a) = (a - r, a + r) \subset (0, 1).$$

Thus a is an interior point of A. Since the choice of a is arbitrary, it follows that every point of A is an interior point and thus A is open.

(b) Let $X = R$ and let $A = (0, 1]$. It can be shown as in the case above that for every $1 \neq a \varepsilon A$, there exists $r > 0$ such that

$$S_r(a) = (a - r, a + r) \subset (0, 1]$$

But there exists no $r > 0$ for which

$$S_r(1) = (1 - r, 1 + r)$$

will lie in $(0, 1]$. Hence every point of A other than 1 is an interior point of A and thus A is not open.

Similarly, if we take $A = [0, 1)$, then every point of A other than 0 is an interior point of A and in the closed interval $[0, 1]$ every point is an interior point, except the end points 0 and 1.

(c) Let $X = R^2$ and let $A = \{x = (x_1, x_2) \varepsilon R^2 : x_1^2 + x_2^2 < 1\}$

Then every point of A is an interior point and hence A is open.

To show this let $x \varepsilon A$. Then $d(x, 0) = x_1^2 + x_2^2 < 1$. So $1 - d(x, 0) = r > 0$. Now

$$S_r(x) = \{y : d(x,y) < r\} \subset A$$

since, if $y \varepsilon S_r(x)$, then $d(x,y) < r$ and so

$$d(y, 0) \le d(y, x) + d(x, 0)$$
$$< r + d(x,0)$$
$$= r + (1 - r) = 1$$

(d) Let $A = R^2$ and

$$A = \{x = (x_1, x_2) \,\varepsilon\, R^2 : x_1^2, x_2^2 \le 1\}$$

This consists of all points within and on the unit circle. Every point within the unit circle is an interior point, whereas the points on the unit circle are not interior points of A and hence A is not open (Fig. 2.7).

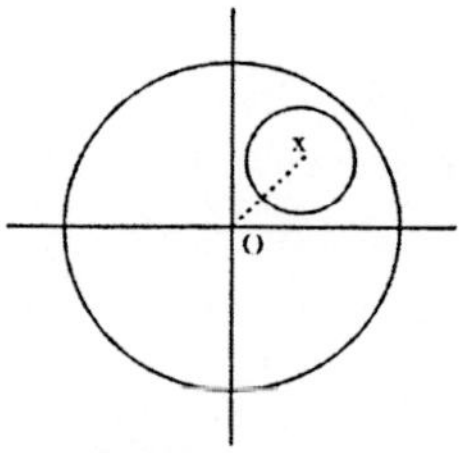

Figure 2.7

(e) The set

$$A = \left\{\frac{1}{n} : n \,\varepsilon\, N\right\}$$

is not open. No point of A is an interior point.

2.2.5. Theorem

Every open sphere in a metric space is an open set.

Proof: Let (X, d) be a metric space. Let $S_r(a)$ be an open sphere in X and let $x \varepsilon S_r(a)$. We have to produce an open sphere with centre x which will be contained in $S_r(a)$. Since $x \varepsilon S_r(a)$, $d(x,a) < r$. Hence $r_1 = r - d(x,a) > 0$ (Fig. 2.8). Now $S_{r_1}(x) \subset S_r(a)$. To show this let $y \varepsilon S_{r_1}(x)$.

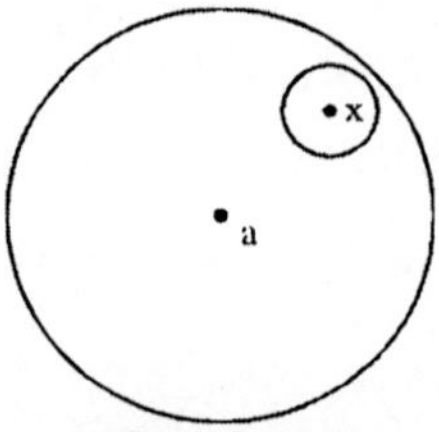

Figure 2.8

Then $d(y,x) < r_1$. Therefore

$$\begin{aligned} d(y,a) &\leq d(y,x) + d(x, a) \\ &< r_1 + d(x, a) \\ &= r - d(x,a) + d(x,a) \\ &= r \end{aligned}$$

Thus $y \in S_r(a)$ and this completes the proof.

2.2.6. Theorem

A subset G in a metric space (X, d) is open iff it is a union of open spheres.

Proof: Let (X, d) be a metric space and let $G \subset X$ be an open set. Then each of its points is the centre of an open sphere contained in G. Hence G is the union of all the open spheres contained in it.

Conversely suppose that G is the union of a collection of open spheres and let $x \in G$. Then x must lie in some open sphere $S_r(a) \subset G$. Since $S_r(a)$ is an open set and $x \in S_r(a)$, there exists an open sphere $S_{r_1}(x)$ contained in $S_r(a)$. Hence $S_{r_1}(x) \subset G$ and thus G is open.

2.2.7. Theorem

Let (X, d) be a metric space. Then

(i) $\varnothing$ and X are open sets,

(ii) the union of any number of open sets in X is open,

(iii) the intersection of a finite number of open sets in X is open.

Proof: (i) Each open sphere centred on each of the points of X is clearly contained in X and hence X is open. The requirement of the definition of an open set is trivially satisfied for the empty set $\varnothing$ and hence $\varnothing$ is open.

(ii) Let $\{G_i : i \in I\}$, where I is any index set, be a family of open sets in X and let $G = \bigcup_{i \in I} G_i$. If $\{G_i\}$ is empty, then G is empty and hence is open. Assume that $\{G_i\}$ is nonempty. Since each G_i is a union of open spheres, G is the union of all such open spheres and hence is open.

(iii) Let $\{G_i : i = 1, 2,...,n\}$ be a finite collection of open sets in X and let $G = \bigcap_{i=1}^{n} G_i$. If $\{G_i\}$ is empty, then $G = X$ and hence is open.

Assume that $\{G_i\}$ is nonempty. If G is empty, then clearly it is open. So we suppose that $G \neq \varnothing$. Let $x \in G$. Then $x \in G_i$ for each $i = 1, 2, \ldots, n$. But each G_i is open. Hence for each i there is $r_i > 0$ such that $S_{r_i}(x) \subset G_i$. Let $r = \min(r_i, r_2, \ldots r_n)$. Then $S_r(x) \subset \bigcap_{i=1}^{n} G_i = G$ and hence G is open.

2.2.8. Remarks

(a) However, it should be noted that the intersection of any number of open sets is not necessarily open. To see this, let $X = R$. Let

$$G_n = \left(\frac{-1}{n}, \frac{1}{n}\right) = \left\{x \in R : -\frac{1}{n} < x < \frac{1}{n}\right\}$$

Then each G_n is an open interval in R and hence is an open set. But in this case

$$\bigcap_{n=1}^{\infty} G_n = \bigcap_{n=1}^{\infty} \left(-\frac{1}{n}, \frac{1}{n}\right) = \{0\}$$

which is not an open set (Fig. 2.9)

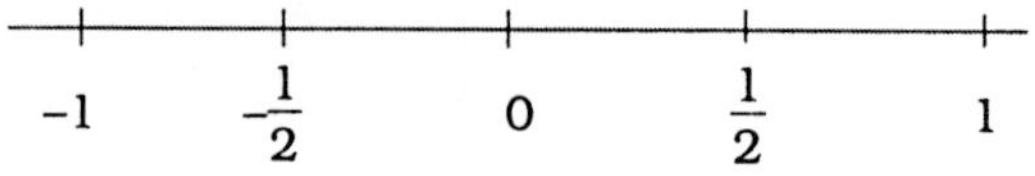

Figure 2.9

(b) In any metric space (X, d) any *singleton set* (i.e., a set consisting of one point only) is not an open set.

We have seen that every open set in a metric space is the union of open spheres. In the case of real line, something more is true: namely it is the countable union of disjoint open spheres.

2.2.9. Theorem

Every open set on the real line is the countable union of disjoint open intervals and such a representation is unique except for the order in which the terms appear in the union.

Proof: Let G be any open set in R. If G consists of a single interval, then there is nothing to prove, If not, let $a \in G$. Since G is open and $a \in G$, there exists an open interval l such that $a \in l \subset G$. Let I_a be the largest open interval with the above property. Such an interval exists, it is the interval having for the left end-point the glb of the left end-points of all these I's and for the right end-point the lub of the right end points of all the I's. It is clear that $G = \underset{a \varepsilon G}{U} I_a$.

Let $b \varepsilon G$. Then we assert that the intervals I_a and I_b are either identical or disjoint. For, otherwise, $a \varepsilon I_a \cup I_b \subset G$, $I_a \subset I_a \cup I_b$ and this contradicts the fact that I_a is the largest interval with $a \varepsilon I_a \subset G$. Thus G is the union of disjoint open intervals.

In order to show that the family of all such intervals is countable, observe that every such interval contains a rational number and distinct intervals contain distinct rationals. Choose one rational from each such distinct interval and let A denote the set of all such rationals. Now $A \subset Q$. But Q is countable and every subset of a countable set is countable. Hence A is countable. Further A is in 1–1 correspondence with the family of disjoint open intervals whose union is G and hence the family of such intervals is countable.

Finally, for the uniqueness assume that there are two different representations of G. Then there exists a point $a \varepsilon G$ which belongs to the interval I of the first representation and to the interval $J \neq I$ of the second representation. Since $J \neq I$, one of them would extend beyond the other. To be precise, let J extend beyond I. If follows from this that one of the end-points of I belongs to J. This is impossible since G is open and end points of I do not belong to G whereas $J \subset G$. This completes the proof of the theorem.

2.3 CLOSED SETS

2.3.1. Definition

Let (X, d) be a metric space and let $A \subset X$. A point $x \varepsilon X$ is called a *limit point* of A if every open ball with centre x contains a point of A other than x. In particular, a point $x \varepsilon R$ is a limit point of $A \subset R$ if, for every $\varepsilon > 0$, $(x - \varepsilon, x + \varepsilon)$ contains a point of A other than x. A point which is not a limit point is called an *isolated point*. A set in a metric space is said to be closed if it contains all its limit points.

2.3.2. Examples

(a) Let $A = \left\{\frac{1}{n} : n \varepsilon N\right\}$. Thus

$A = \left\{1, \frac{1}{2}, \frac{1}{3}, \ldots\right\} \subset Q$. The only limit point of A is 0 which is not a member of the set. Hence A is not closed.

(b) Let $A = (0, 1) \subset R$.

Every point of A is a limit point of A. Further 0 and 1 are also limit points of A but they are not in A. Thus A is not closed.

(c) Let $A = (0, 1) \subset R$.

In this case every point of A is a limit point of A and A has no other limit point. Thus A is a closed set.

(d) Let $A = \{x = (x_1, x_2) \, \varepsilon \, R^2 : x_1^2 + x_2^2 < 1\}$

Thus A consists of all points inside the unit circle with centre at the origin. In this case every point of A is a limit point of A, moreover every point on the unit circle is also a limit point.

2.3.3. Theorem

Let (X, d) be a metric space. Let $A \subset X$ and let x be a limit point of A. Then every neighbourhood of x contains infinitely many points of A.

Proof: Let us assume, to the contrary, that there is a neighbourhood $S_\epsilon(x)$ of x which contains only finite number of points of A. Let $x_1, x_2, \ldots, x_n$ be those points of A which are different from x and lie in $S_\epsilon(x)$. Let

$$r = \min_{m = 1, 2, \ldots, n} d(x, x_m)$$

Then $r > 0$ and the neighbourhood $S_r(x)$ contains no point of A other than x. So x cannot be a limit point of A and this contradiction proves the result.

2.3.4. Corollary

A finite set in any metric space has no limit point.

2.3.5. Theorem

A subset A of a metric space (X, d) is closed iff its complement is open.

Proof: Assume that A is closed. We shall show that A^c is open. Let $x \, \varepsilon \, A^c$. Then there exists an $r > 0$ such that , $S_r(x) \subset A^c$, for, if this is false, then for every $r > 0$ $S_r(x)$ would contain points of A and thus x would be a limit point of A. Since A is closed, in that case, $x \in A$. This contradicts the assumption that $x \in A^c$ and hence A^c is open.

Conversely, assume that A^c is open and let x be a limit point of A. Now $x \in A$. For if $x \in A^c$, then since A^c is open, x is an interior point of A^c. So there is an $r > 0$ such that $S_r(x) \subset A^c$. Thus $S_r(x)$ contains no point of A and hence x cannot be a limit point of A. Since by assumption x is an arbitrary limit point of A, it follows that A contains all its limit points and hence A must be closed.

2.3.6. Definition

Let (X, d) be a metric space. Let $x \in X$ and $r > 0$. The closed sphere or closed ball of radius r with centre x is defined by

$$S_r[x] = \{y \varepsilon X : d(x,y) \leq r\}$$

2.3.7. Theorem

Every closed sphere in a metric space is a closed set.

Proof: Let (X,d) be a metric space, $x \in X$ and $r > 0$. We shall show that $(S_r[x])^c$ is open, or, equivalently, that every point of $(S_r[x])^c$ is an interior point. Let $y \in (S_r[x])^c$. Then $y \in (S_r[x])$ and hence $d(x,y) > r$. Let $r_1 = d(x,y) - r > 0$. We now claim that $S_{r_1}(y) \subset (S_r[x])^c$. To show this, let $z \in S_{r_1}(y)$. Then $d(y, z) < r_1$. Now by triangle inequality

$$d(x, y) \leq d(x, z) + d(z, y)$$
$$< r_1 + d(x, z)$$

Hence,

$$d(x, z) > d(x, y) - r_1 = r$$

So $z \notin S_r[x]$. Therefore $z \in (S_r[x])^c$ and this completes the proof.

2.3.8. Theorem

Let (X, d) be a metric space. Then

(i) $\emptyset$ and X are closed sets,

(ii) the union of a finite collection of closed sets is closed,

(iii) the intersection of an arbitrary collection of closed sets is closed.

Proof: The proof uses Theorem 2.2.7 and DeMorgan's laws.

(i) Since X and $\emptyset$ are open sets, $X = \emptyset^c$ and $\emptyset = X^c$ are closed sets.

(ii) Let $A_1, A_2, \ldots A_n$ be a finite collection of closed sets. Then

$$\left(\bigcup_{i=1}^{n} A_i\right)^c = \bigcap_{i=1}^{n} A_i^c$$

Since each A_i is closed A_i^c is open and so is their intersection. Hence $\bigcup_{i=1}^{n} A_i$ is closed.

(iii) Let $\{A_i \ : i \varepsilon I\}$ be an arbitrary collection of closed sets. We have

$$\left(\bigcap_{i\in 1} A_i\right)^c = \bigcup_{i\in 1} A_i^c$$

Since each A_i is closed, A_i^c is open. The set $\bigcup_{i\in 1} A_i^c$, being the union of open sets, is open. Thus $\left(\bigcap_{i\in 1} A_i\right)^c$ is open and hence $\bigcap_{i\in 1} A_i$ is closed. This completes the proof.

It may, however, be noted that a union of an arbitrary collection of closed sets may not be closed. For example, let $X = R$ and $A_n = \left[\frac{1}{n}, 2\right]$ *for* $n \in N$. Then each A_n is closed but

$$\bigcup_{n=1}^{\infty} A_n = \bigcup_{n=1}^{\infty} \left[\frac{1}{n}, 2\right] = (0, 2)$$

which is not closed.

We now give a very important example of a peculiar closed set which is very useful in many connections, particularly in the theory of integration.

2.3.4. Cantor's Ternary Set

Let $I = [0, 1]$. Divide I into three equal parts and remove the open middle part $I_1 = \left(\frac{1}{3}, \frac{2}{3}\right)$ to get a set

$$I - I_1 = \left[0, \frac{1}{3}\right] \cup \left[\frac{2}{3}, 1\right]$$

Now divide each of the intervals $\left[0, \frac{1}{3}\right]$ and $\left[\frac{2}{3}, 1\right]$ into three parts and remove the open middle parts

$$I_2 = \left(\frac{1}{3^2}\right)\left(\frac{2}{3^2}\right) \cup \left(\frac{7}{3^2}\right)\left(\frac{8}{3^2}\right)$$

to get the set

$$I - (I_1 \cup I_2) = \left[0, \frac{1}{3^2}\right], \cup \left[\frac{2}{3^2}, \frac{3}{3^2}\right] \cup \left[\frac{6}{3^2}, \frac{7}{3^2}\right], \cup \left[\frac{8}{3^2}, \frac{9}{3^2}\right]$$

Let I_3 be the union of the middle one-thirds of each of the above intervals. We remove this to get the set

$$I - \bigcup_{n=1}^{3} I_n$$

and we continue this process. The resulting set we get is

$$P = I - \bigcup_{n=1}^{\infty} I_n = \left(\bigcup_{n=1}^{\infty} I_n\right)^c$$

This is called Cantor's ternary set. Since each I_n is open so is their union. Hence P, being the complement of an open set, is closed.

2.3.10. Definition

Let (X, d) be a metric space and let $A \subset X$.

(a) A is said to be *prefect* if A is closed and every point of A is a limit point of A,

(b) The set of all limit points of A is called the *derived set* of A and is denoted byA',

(c) A is said to be *dense* in X if $X = A \cup A'$, that is, if every point of X is either a limit point of A or is a point of A,

(d) The smallest closed set containing a is called the *closure* of *A and is denoted by* $\overline{A}$.

2.3.11. Examples

(a) If $A = \left\{\frac{1}{n} : n \in N\right\}$, then $A' = \{0\}$.

(b) If $A = [0, 1] \subset R$ then $A' = [0, 1]$ and $\overline{A} = [0, 1]$.

(c) If $A = [0, 1] \subset R$, then

$A' = [0, 1] = A$ and $\bar{A} = [0, 1] = A$.

Thus A is perfect.

(d) Let $A = \{x = (x_1, x_2) \in R^2 : x_1^2 + x_2^2 < 1\}$

Then $A' = \{x = (x_1, x_2) \in R^2 : x_1^2 + x_2^2 \leq 1\} = \bar{A}$

(e) Let $A = \{x = (x_1, x_2) \in R^2 : x_1^2 + x_2^2 \leq 1\}$

In this case every point of A is a limit point of A and A has no other limit point. Thus $A' = A = \bar{A}$. A is perfect.

2.3.12. Theorem

Let (X,d) be a metric space and let $A \subset X$. Then $\bar{A} = A \cup A'$. In other words, closure of A is the union of points of A and limit points of A.

Proof: We shall first show that $A \cup A' \subset \bar{A}$. Since $A \subset \bar{A}$ it is, therefore, enough to show that $A' \subset \bar{A}$. For this let $x \notin \bar{A}$. Then $x \in (\bar{A})^c$. Since $\bar{A}$ is closed, $(\bar{A})^c$ is an open set. Therefore x is an interior point of $(\bar{A})^c$, that is, there exists some $r > 0$ such that

$$S_r(x) \subset (\bar{A})^c \subset A^c$$

Thus it follows that $S_r(x)$ does not contain any point of A and hence x cannot be a limit point of A. Thus $x \notin A'$. Hence

$$A' \subset \bar{A} \text{ and so } A \cup A' \subset \bar{A}$$

Conversely, let $x \in \bar{A}$, if $x \in A$ then $x \in A \cup A'$ and hence $\bar{A} \subset A \cup A'$. If $x \notin A$, then we shall show that $x \in A'$, i.e., that x is a limit point of A. Assume, to the contrary, that it is not so. Then there exists some $r > 0$ such that $S_r(x)$ contains no point of A and so

$$S_r(x) \subset A^c$$

So

$$(S_r(x))^c \supset (A^c)^c = A.$$

Since $S_r(x)$ is open, $(S_r(x))^c$ is closed and $(S_r(x))^c \supset (A^c)^c$.

Therefore,

$$(S_r(x))^c \supset \bar{A}$$

But $x \in \bar{A}$ and hence $x \in (S_r(x))^c$. But this is impossible. Hence $x \in A'$ and so

$$\bar{A} \subset A \cup A'$$

Thus $\bar{A} = A \cup A'$ and this completes the proof.

2.3.13. Remark

From the above theorem it follows that the definition of *denseness* of a set can be reformulated in the following way : If A is a subset in a metric space X, then A is said to be dense in X if $\bar{A} = X$.

2.3.14. Theorem (Cantor's Intersection)

Let $I_1, I_2, \ldots$ be intervals in R such that $I_{n+1} \subset I_n$ for each $n \in N$. Then

$$\bigcap_{n=1}^{\infty} I_n \neq \varnothing$$

Proof: Let $I_n = [a_n, b_n]$. Let $X = \{a_n : n \in N\}$

Then $X \neq \varnothing$ and is bounded above by b_1. Therefore X has a lub. Let x = lub of X. If m and n are positive integers, then

$$a_n \leq a_{m+n} \leq b_{m+n} \leq b_m$$

so that $x \leq b_m$ for each m. Obviously $a_m \leq x$ and hence $x \in I_m$ *for* all $m \in N$. Thus $x \in \bigcup_{m=1}^{\infty} I_m$ and this completes the proof.

2.4. COMPACTNESS

2.4.1. Definition

Let (X, d) be a metric space. A family $G = \{G_i : i \in I\}$ of subsets of X is called a cover of X if $\bigcup_{i \in I} G_i = X$. If each G_i is open, then the cover $G = \{G_i : i \in I\}$ is called an open cover. If the index set I is finite, then G is called a finite cover. A subfamily of a cover which is also a cover is called a subcover of the original cover.

2.4.2. Example

Let $A_n = (n, n+3)$ for $n = 1, 2, \dots$ Then $G = \{A_n : n \varepsilon Z\}$ is an open cover of the whole real line R. The family

$$\{A_n : n \text{ even}\}$$

is also an open cover for R, but this is a subfamily of G and hence is a subcover of G.

2.4.3. Definition

A metric space (X, d) is said to be *compact* if every open cover of it has a finite subcover.

Thus if X is compact and if $G = \{G_i : i \in I\}$ is any family of open subsets of X satisfying $\bigcup_{i \in I} G_i = X$, then there exists a finite subfamily $G_{i_1}, G_{i_2}, \dots, G_{i_n}$ of G such that

$$G_{i_1} \cup G_{i_2} \cup \dots \cup G_{i_n} = X$$

Note that the whole real line R is not compact. Because the family

$$\{(-n, n) : n \varepsilon Z\}$$

is an open cover of R but it does not have a finite subcover.

We shall now show that every closed and bounded interval on the real line R is compact. This is a particular case of a famous theorem known as the Heine-Borel Theorem. This theorem, in its most general form, states that a subset of R^n is compact iff it is closed and bounded and this has far-reaching applications. The particular case which we will prove now is also called the Heine-Borel Theorem.

2.4.4. Heine–Borel Theorem

Any closed interval $[a, b] \subset R$ is compact.

Proof: Let $G = \{G_i : i \in I\}$ be an open cover of $[a,b]$. Let C denote the set of all points $c \in [a, b]$ satisfying the condition that $[a, c]$ is covered by a finite subcollection of open sets of G. Observe that $a \in C$ and hence $C \neq \varnothing$. Further, if $c \in C$ and $z \in [a, c]$, then $[a,z]$ is covered by a finite sub-collection of open sets of G, so $z \in C$, and hence the entire interval $[a, c] \subset C$. Thus the set C is an interval of the form $[a,x]$. We claim that $x = b$. Assume, to the contrary, that $x \neq b$, then clearly $x < b$. Since $x \in G_i$ for some i and each G_i is open, there is an interval $(x - \in, x + \in) \subset G_i$. The fact that $[a,x]$ is covered by a finite number of open sets in G implies that $[a, x + \frac{\in}{2}]$ is also covered by a finite number of open sets in G. We have thus a contradiction and this completes the proof.

2.4.5. Theorem

Every infinite subset A of a compact metric space (X, d) has a limit point in X.

Proof: Let X be a compact metric space and $A \subset X$ be an infinite set. *We have to show that* A has limit point in X. Assume to the contrary that no point of X is a limit point of A. Then, for each $x \in X$ there exists an $r_x > 0$ such that $S_{r_x}(x)$ contains no point of A other than x (if $x \in A$). Thus the open cover$\{S_{r_x}(x) : x \in X\}$ has no finite subcover for A and hence has no finite subcover for X since $A \subset X$. This contradicts the assumption that X is compact and hence the result follows.

The famous Bolzano-Weierstrass Theorem is now obtained from above two results.

2.4.6. Bolzano–Weierstrass Theorem

Every bounded infinite subset of R has a limit point.

Proof: Let A be a bounded infinite set in R. Then there exists a closed bounded interval, say $[a,b]$ in R which contains A. By the Heine - Borel Theorem, $[a,b]$ is compact. Thus A is an infinite subset of the compact set $[a,b]$. therefore A has limit point and this completes the proof.

2.5 CONNECTEDNESS

2.5.1. Definition

A metric space (X, d) is said to be *disconnected* if there exist two disjoint, nonempty open sets A and B of X such that $X = A \cup B$. A representation of a disconnected space X in the form $X = A \cup B$ as above is called a *disconnection* of X. A space X is said to be connected if it is not disconnected.

2.5.2. Remark

(a) If X is connected then it cannot be expressed as $X = A \cup B$, where A and B are disjoint, open and nonempty. In other words, if X is connected such that $X = A \cup B$, where A and B are disjoint and open, then either A or B must be empty.

(b) Let $X = A \cup B$ be a disconnection of X. Then $A^C = B$ and $B^C = A$. Again A and B are both open, hence A and B are both closed also. In other words, if X is disconnected, then it can be expressed as the union of two disjoint, nonempty sets which are both open and both closed.

(c) If $X = A \cup B$ where A and B are disjoint and nonempty and if X is connected, then one of A and B is open and the other is closed.

2.5.3. Examples

Let $X = (0, 1) \cup (2, 3)$. Then X is disconnected. Let $1 < a < 2$. We have

$$(0, 1) = (-\infty, a) \cap X$$

$$(2, 3) = (a, \infty) \cap X$$

Now $(0, 1) \neq \varnothing$, $(2, 3) \neq \varnothing$, $(0, 1) \cap (2, 3) = \varnothing$. $(0, 1)$ and $(2, 3)$ are both open and being complements of open sets are both closed. Thus $X = [(-\infty, a) \cap X] \cup [(a, \infty) \cap X]$ is a disconnection of X.

We shall now show that a subset of the real line R is connected iff it is an interval.

2.5.4. Theorem

A subset of R is connected iff it is an interval.

Proof: Let $X \subset R$ be connected. We have to show that X is an interval. Assume, to the contrary, that it is not. Then there exist $x, y, z \in R$ such that $x < y < z$, x and z are in X but $y \notin X$.

Then

$$x = [(-\infty, y) \cap X] \cup [X \cap (y, \infty)]$$

is a disconnection of X. This contradicts the fact that X is connected.

Conversely, assume that X is disconnected. We shall show that it is not an interval. Let $X = A \cup B$ be a disconnection of X. A and B are nonempty and disjoint. So we can choose points $x \in A$ and $z \in B$ such that $x \neq z$. We can assume, without any loss of generality, that

$x < z$. Assume that X is an interval. Then $[x, z] \subset X$ and each point of $[x, z]$ is either in A or in B.

Let

$$y = \sup ([x, z] \cap A)$$

Clearly $x \leq y \leq z$ *and hence* $y \in X$. Since A is closed, $y \in A$ and hence $y < z$. Again, from the definition of y it follows that $y + \in$ is in B for every $\in > 0$ satisfying $y + \varepsilon \leq z$. Now since B is closed, $y\ \varepsilon\ B$. Hence $y\ \varepsilon\ A \cap B$ and this contradicts the fact that $A \cap B = \emptyset$. This contradiction proves the result.

2.5.5. Corollary

R is connected.

EXERCISES

1. Answer 'Yes' or 'No'.
   (i) The set R is a metric space with $d(x, y) = |x-y|$ for $x, y\ \varepsilon\ R$.
   (ii) The set $\emptyset$ is open in a metric space (X, d).
   (iii) The intersection of two open sets is open.
   (iv) A closed set in (X, d) does not have an interior point.
   (v) In the metric space (X, d), the set X cannot be an ordered field.
   (vi) In (X, d) the sphere $\{y : d(x, y) \leq r\}$ is a closed set.
2. Correct (if necessary) the following statements;
   (i) An infinite set in a metric space must have atleast one limit point.
   (ii) an isolated point of a set may have a neighbourhood containing points of the set.
   (iii) In a metric space (X, d) any union of closed sets is closed.
   (iv) A closed interval on a real line is a compact set.
   (v) If A and B are connected subsets of a metric space (X, d) then

$$A \cap B = \emptyset \Rightarrow A \cup B \text{ is connected.}$$

3. State whether true or false:
   (i) $A = (a, b)$, $A' = \{a, b\}$ and $\bar{A} = [a, b]$.
   (ii) For any $A \subset Q$, $A \neq \bar{A}$.
   (iii) $[a, b] \times (c, d)$ is compact.

(iv) A^c is compact where
$A = \{x \in R \mid 0 < x < 1\}$

4. For $x, y \in R$ define

$$d_1(x, y) = (x - y)^2$$

$$d_2(x, y) = \sqrt{|x - y|},$$

$$d_3(x, y) = |x^2 - y^2|$$

Determine, for each of these, whether it is a metric on R.

5. Let $X = R$ and define for $x, y \in R$.

$$d(x, y) = |x - y|^{\alpha}, \alpha \in R \text{ and } 0 < \alpha < 1.$$

Show that (R, d) is a metric space.

6. Let (X, d) be a metric space. Define

$$d_1(x, y) = \min [d(x,y), 1]$$

Show that (X, d_1) is also a metric space.

7. If (X, d) is a metric space and if $x, y, z \in X$, then show that

$$|d(x, z) - d(y, z)| \leq d(x, y)$$

8. If (X, d) is a metric space and if $x, y, x', y' \in X$, then show that

$$|d(x,y) - d(x',y')| \leq d(x,x') + d(y,y')$$

9. Let Int A denote the set of all interior points of a set A in a metric space (X, d). If $A \subset X$ and $B \subset X$, then show that

(i) Int $(A \cap B)$ = Int $A \cap$ Int B

(ii) Int $(A \cup B)$ = Int $A \cup$ Int B.

10. Find the interiors of the following subsets of R^2 in the standard metric.

(i) $A = \{(x_1, x_2) \in R^2 : x_1^2 + x_2^2 \leq 1\}$

(ii) $B = \{(x_1, x_2) \in R^2 : x_1^2 + x_2^2 < 1\}$

(iii) $C = \{(x_1, x_2) \in R^2 : x_1^2 + x_2^2 = 1\}$

(iv) $D = \{(x_1, x_2) \in R^2 : x_2 \geq 0\}$

11. The closure of a set in a metric space is the smallest closed set (the intersection of all closed sets) containing the set. Find the closure of the following subsets of R in the standard metric.

(i) $A = (0, 1) \cup (1, 2) \cup (2, 3)$

(ii) $B = [0, 1) \cup (1, 2]$

(iii) $C \neq (0, 1) \cup (2, 3)$

12. We write $\overline{A}$ to denote the closure of a set in a metric space. If A_1, A_2, ..., A_n are subsets of a metric space, then show that

(i) $B_n = \bigcup_{i=1}^{n} A_i \Rightarrow \overline{B}_n = \bigcup_{i=1}^{n} \overline{A}_i$

(ii) $B = \bigcup_{i=1}^{\infty} A_i \Rightarrow \overline{B} \supset \bigcup_{i=1}^{\infty} \overline{A}_i$

13. The set of all limit points of a set A in a metric space is called the derived set of A. Give an example of a set whose derived set is

(i) $\varnothing$

(ii) a subset of the given set,

(iii) a superset of the given set,

(iv) neither a subset nor a superset of the given set,

(v) the same as the set.

14. We write A' to denote the derived set of A. Show that

(i) $A \subset B \Rightarrow A' \subset B'$

(ii) $(A \cup B)' = A' \cup B'$

(iii) $(A \cap B)' \subset A' \cap B'$

15. A point of a subset A in a metric space is called the boundary point if it is not an interior point. Find the boundary points of the following subsets of R^2 in standard metric.

(i) $A = \{ (x_1, x_2) \in R^2 : x_1^2 + x_2^2 \leq 1\}$

(ii) $B = \{ (x_1, x_2) \in R^2 : x_1^2 + x_2^2 < 1\}$

(iii) $C = \{ (x_1, x_2) \in R^2 : x_1^2 + x_2^2 = 1\}$

3 Sequences and Series

3.1. SEQUENCES IN A METRIC SPACE

3.1.1. Definition

Let (X, d) be a metric space. A *sequence* in X is a mapping x from N into X, thus a *real sequence* is a map from N into R and a *complex sequence* is a map from N into C. We write x_n to denote the value of x at $n \in N$. A sequence x is denoted by $\{x_n\}$ or $(x_1, x_2, \ldots x_n, \ldots)$ or by $x = \{x_n\}$. The set $\{x_n : n \in N\}$ is called the range of the sequence. The range of a sequence may be a finite set, since a sequence need not be a 1–1 map, it may be possible that it maps N into a finite set. In the following examples we consider only real sequences.

3.1.2. Examples

(a) Let $x_n = (-1)^n$. Then

$$x = \{x_n\} = (-1, 1, -1, 1, \ldots)$$

Thus the range of $x = \{1, -1\}$

(b) If $x = (1, 1, 1, \ldots)$ then the range of $x = \{1\}$.

(c) The range of

$$x = \{x_n\} \text{ where } x_n = \frac{1}{n}$$

is an infinite set.

(d) If $x = \{x_n\} = \left\{\frac{(-1)^n}{n}\right\}$, then the range of x is also infinite.

(e) The range of $\left\{\frac{1-(-1)^n}{2}\right\}$ is the set $\{0, 1\}$.

(f) The range of $\{\sin n\pi\}$ is $\{0\}$.

(g) The range of $\{n\} = (1, 2, \ldots)$ is the set N.

We write e to denote the sequence (1, 1,...) and 0 to denote the sequence (0, 0...).

3.1.3. Definition

A sequence is said to be bounded if its range is a bounded set. More precisely, a sequence x in (X, d) is bounded if there exists a constant $K > 0$ such that

$$d(x_n, 0) \leq K \text{ for all } n \in N.$$

If $X = R$, then the condition reads as

$$|x_n| \leq K \text{ for all } n \in N.$$

If a real sequence $\{x_n\}$ is bounded, then it follows from the order completeness of R that $\{|x| : n \in N\}$ has a lub or supremum and we write it as sup $|x_n|$.

3.1.4. Examples

(a) The sequences in Examples 3.1.2 (a), (b), (c), (d), (e) and (f) are bounded.

(b) The sequence $\{n\}$ is not bounded. Similarly the sequence $\{(-1)^n n\}$ is not bounded.

3.1.5. Definition

A sequence $\{x_n\}$ in a metric space (X, d) is said to be convergent to a limit $s \in X$ if for every $\varepsilon > 0$ there exists an $n_0 \in N$ such that

$$d(x_n, s) < \epsilon \text{ for } n \geq n_0.$$

In other words, if for every $\epsilon > 0$ all but a finite number of terms of the sequence lie within the open ball of centre s and radius ϵ. In particular, a real sequence $\{x_n\}$ converges to a point $s \in R$ if for every $\varepsilon > 0$ there exists a positive integer n_0 such that for all $n \geq n_0$,

$|x_n - s| < \varepsilon$

i.e., $-\varepsilon < x_n - s < \varepsilon$

i.e., $s - \varepsilon < x_n < s + \varepsilon$.

This means that, for every $\varepsilon > 0$, all but a finite number of terms of the sequence lie in the open interval $(s - \varepsilon, s + \varepsilon)$. A sequence which does not converge is said to be divergent. If a sequence $\{x_n\}$ converges to s we write

$$\lim_{n\to\infty} x_n = s \text{ or } x_n \to s \text{ as } n \to \infty$$

The convergence of a sequence does not depend only on the sequence itselt, it also depends on the metric space, and this will be clear from the following examples.

3.1.6. Examples

(a) The sequence $x = \left\{\frac{1}{n}\right\}$ is convergent to 0 if we consider the metric space X to be R or [0, 1} or any subset of R which contains [0, 1], but it does not converge if $X = (0, 1]$.

(b) The sequence e converges to 1, {0} converges to 0.

(c) The sequences $\left\{(-1)^n, \frac{1-(-1)^n}{2}\right\}$ are not convergent.

(d) The sequences $\{n\}$, $\{n^2\}$, $\{-1)^n n\}$ are not convergent.

3.1.7. Theorem

Every convergent sequence in a metric space has a unique limit.

Proof: Let (X, d) be a metric space let $\{x_n\}$ be a convergent sequence in X which converges to $s \varepsilon X$. Assume, if possible, that $\{x_n\}$ converges to $s \varepsilon X$. Let $\varepsilon > 0$ be given. Then there exist n_1 and n_2 such that

$$d(x, s) < \frac{\varepsilon}{2} \text{ for } n \geq n_1$$

$$d(x_n, t) < \frac{\varepsilon}{2} \text{ for } n \geq n_2.$$

Let $n_0 = \max(n_1, n_2)$. Then for $n \geq n_0$,

$$d(s, t) \leq d(x_n, s) + d(x_n, t)$$

$$< \frac{\varepsilon}{2} + \frac{\varepsilon}{2} = \varepsilon$$

Since ε is arbitrary, we must have $d(s, t) = 0$. Thus $s = t$ and this completes the proof.

3.1.8. Theorem

Every convergent sequence in a metric space is bounded, but the converse is not true.

Proof: Let (X, d) be a metric space and let $\{x_n\}$ be a sequence in X which converges to $s \varepsilon X$. We have to show that $\{x_n\}$ is bounded, in other words, we have to show that there exists a constant $K > 0$ such that

$$d(x_n, s) \leq K \text{ for all } n \varepsilon N.$$

Let $\varepsilon > 0$ be given. Since $\{x_n\}$ conveges to $s \varepsilon X$, there exists $n_0 \varepsilon N$ such that

$$d(x_n, s) < \varepsilon \text{ for } n \geq n_0$$

Therfore for $n \geq n_0$,

$$d(x_n, 0) \leq d(x_n, s) + d(s, 0) < \varepsilon + d(s, 0).$$

Let

$$M = \max\{d(x_1, 0), d(x_{2,0}), \ldots, d(x_{n_0-1}, 0)\}$$

and

$$K = \max\{M, \varepsilon + d(s, 0)\}$$

Then it follows that

$$d(x_n, 0) \leq K \text{ for all } n.$$

To see that the converse is not true observe that the sequences $\{(-1)^n\}$, $\left\{\frac{1-(-1)^n}{2}\right\}$ are bounded but not convergent.

The following theorem gives the relation between the limit of a convergent sequence and the limit point of the range of the sequence.

3.1.9. Theorem

(a) If the range of a convergent sequence in a metric space is an infinite set, then the limit of the sequence is a limit point of the range.

(b) If s is a limit point of a set E in a metric space, then there is a convergent sequence $\{x_n\}$ in E with limit s.

Proof: (a) Let (X, d) be a metric space and let $\{x_n\}$ be a sequence in X with limit $s \varepsilon X$. Assume that x is not a limit point of the range $\{x_n : n \varepsilon N\}$. Then there exists some $r > 0$ such that the open ball $S_r(s)$ does not contain any point of the sequence other than s. But s is the limit of the sequence. Hence all the terms of $\{x_n\}$, except a finite number of them, will lie in $S_r(s)$ and therefore must coincide with s. Thus there are only finitely many distinct points in the sequence. This contradicts the fact that the range of the sequence is infinite and completes the proof.

(b) Let (X, d) be a metric space and $E \subset X$. Let s be a limit point of E. Then for every $n \varepsilon N$ there is a point $x_n \varepsilon E$ such that $d(x_n, s) < \frac{1}{n}$.

Let $\varepsilon > 0$ be given. Choose $n_0 \varepsilon N$ such that $\frac{1}{n_o} < \varepsilon$. Then for $n > n_0$,

$$d(x_n, s) < \frac{1}{n_0} < \varepsilon$$

This shows that the sequence $\{x_n\}$ converges to s and this completes the proof.

3.2 CAUCHY SEQUENCES

3.2.1. Definition

Let (X, d) be a metric space. A sequence $\{x_n\}$ in X is said to be a Cauchy sequence if for every $\varepsilon > 0$ there exists a positive integer n_0 such that

$$d(x_n, x_m) < \varepsilon \text{ for } n, m \geq n_0$$

In particular, a sequence $\{x_n\}$ in R is a Cauchy sequence if for every $\varepsilon > 0$ there exists an $n_0 \varepsilon N$ such that

$$|x_n - x_m| < \varepsilon \text{ for } n, m \geq n_0$$

In other words, a sequence is Cauchy if from and after certain terms any two terms of the sequence are sufficiently close to each other.

3.2.2. Theorem

Every convergent sequence in a metric space is a Cauchy sequence, but the converse is not true.

Proof: Let (X, d) be a metric space and let $\{x_n\}$ be a sequence in X which converges to $s \varepsilon X$. Let $\varepsilon > 0$ be given. Then there exists $n_0 \varepsilon N$ such that

$$d(x_n, s) < \frac{\varepsilon}{2} \text{ for } n \geq n_0.$$

Let $n, m \geq n_0$. Then

$$d(x_n, x_m) \leq d(x_n, s) + d(x_m, s)$$

$$< \frac{\varepsilon}{2} + \frac{\varepsilon}{2} = \varepsilon.$$

This shows that $\{x_n\}$ is a Cauchy sequence.

To show that the converse is not true, take $X = (0, 1)$ and $x_n = \frac{1}{n}$ for all $n \varepsilon N$. then the sequence $\{x_n\}$ is Cauchy, but it does not converge in X. If we take $X = [0, 1]$, then the sequence, of course, converges to 0 (in X).

As another example, take $X = Q$ with the metric d defined by d $(x,y) = |x - y|$ for $x, y \in Q$. Then the sequence

$$\{1, 1.4, 1.414, 1.4146, ...\}$$

is a Cauchy sequence, but it does not converge in Q. If we take $X = R$ and consider the above sequence as a sequence in R, then it of course converges to $\sqrt{2}$ (in R).

It now follows from above examples that every Cauchy sequence in any metric space need not be convergent. Those metric spaces in which this property holds bear a special name and this is given in the following definition.

3.3 COMPLETE METRIC SPACE

3.3.1. Definition

A metric space is said to be complete if every Cauchy sequence in it converges to an element of it. A metric space which is not complete is called incomplete.

3.3.2. Examples

(a) The real line R with the standard metric is a complete metric space where as the set of all rational numbers, Q, it incomplete. On the axiomatic approach to the real line the completeness of R is an axiom which is known as the completeness axiom for the real numbers. If a constructive approach to the reals is taken, starting with the incomplete metric space Q, a new metric space is constructed and its completeness is proved. The complete metric space so obtained is called the completion of Q and is defined as the real line R. We shall, however, not discuss these constructions in this introductory book. Because of the completeness axiom of R and Theorem 3.2.2 it follows that a sequence $\{x_n\}$ of real numbers is convergent if and only if it is Cauchy.

(b) C, The set of all complex numbers is a complete metric space. To prove this, let $\{z_n\}$ be a Cauchy sequence of complex numbers. Let $\varepsilon > 0$ be given. Then there is a positive integer n_0 such that

$$|z_n - z_m| < \varepsilon \text{ for } n, m \geq n_0.$$

Let $z_n = x_n + iy_n$. Then for any two positive integers m and n,

$$|x_n - x_m| \leq |z_n - z_m|$$

and

$$|y_n - y_m| \leq |z_n - z_m|$$

Therefore for $n, m, \geq n_0$ we have

$$|x_n - x_m| < \varepsilon$$

and

$$|y_n - y_m| < \varepsilon$$

This shows that $\{x_n\}$ and $\{y_n\}$ are Cauchy sequences of real numbers. Since R is complete, $\{x_n\}$ and $\{y_n\}$ must respectively converge to some s and t. Now it follows that the sequence $\{z_n\}$ converges to $s + it$. Since $\{x_n\}$ and $\{y_n\}$ converge to s and t, there exist n_1 and n_2 such that

$$|x_n - s| < \frac{\varepsilon}{2} \text{ for } n \geq n_1 \text{ and}$$

$$|y_n - t| < \frac{\varepsilon}{2} \text{ for } n \geq n_2$$

Put $n_0 = \max(n_1, n_2)$. Then for $n \geq n_0$,

$$\begin{aligned} |(x_n + iy_n) - (s + it)| &= |(x_n - s) + i(y_n - t)| \\ &\leq |x_n - s| + |y_n - t| \\ &< \frac{\varepsilon}{2} + \frac{\varepsilon}{2} = \varepsilon \end{aligned}$$

and this completes the proof.

3.4 SEQUENCES OF REAL NUMBERS

In this section we shall discuss sequences of real numbers. We write S to denote the set of all such sequences and define algebraic operations of addition, scalar multiplication, and multiplication in the set S.

3.4.1. Definition

Let $x = \{x_n\}$ and $y = \{y_n\}$ be any two elements of S and let $\alpha \in R$. Then we define new sequences $x + y$, αx and xy by

$$x + y = \{x_n + y_n\}$$
$$\alpha x = \{\alpha \; x_n\}$$
$$xy = \{x_n y_n\}$$
$$-x = \{-x_n\},$$
$$x - y = x + (-y) = \{x_n - y_n\}.$$

If $y_n \neq 0$ for any $n \in N$, then we define

$$\frac{1}{y} = \left\{\frac{1}{y_n}\right\} \text{ and } \frac{x}{y} = \left\{\frac{x_n}{y_n}\right\}$$

3.4.2. Theorem

If $x = \{x_n\}$, $y = \{y_n\}$ and $z = \{z_n\}$ are in S and if $\alpha, \beta \in R$ then

(i) $x + y = y + x$
(ii) $(x + y) + z = x + (y + z)$
(iii) $x + \{0\} = x$
(iv) $x + (-x) = \{0\}$
(v) $\alpha (x + y) = \alpha x + \alpha y$
(vi) $(\alpha + \beta) x = \alpha x + \beta x$
(vii) $(\alpha\beta) x = \alpha(\beta x)$
(viii) $1x = x$

The proof of this is straightforward from the definition and hence is left to the readers as an exercise. Those who know the definition of Vector space will observe that Theorem 3.4.2 just says that S is a vector space over the field of real numbers, The following properties are also true whose easy proof is left to the reader as an exercise.

3.4.3. Theorem

For $x, y, z \in S$ and $\alpha \in R$,

(i) $xy = yx$
(ii) $(xy) z = x(yz)$

(iii) $(x + y)\, z = xz + yz,$

$x(y + z) = xy + xz$

(iv) $\alpha\, (xy) = (\alpha x)\, y$

3.4.4. Theorem

Let $x = \{x_n\}$ and $y = \{y_n\}$ be any two bounded real sequences and let $\alpha \in R$. Then the sequences $x + y$ and αx and xy are all bounded.

Proof: Since x and y are bounded there exist constants $K_1 > 0$ and $K_2 > 0$ such that for all $n \in N$,

$$|x_n| \leq K_1 \text{ and } |y_n| \leq K_2$$

Hence $|x_n + y_n| \leq |x_n| + |y_n|$

$$\leq K_1 + K_2 \text{ for all } n \in N$$

and thus $x + y$ is bounded. Similarly for all $n \in N$,

$$|\alpha x_n| = |\alpha|\ |x_n| \leq |\alpha|\ K_1$$

which shows that αx is bounded. Also

$$|x_n y_n| = |x_n|\ |y_n| \leq K_1\ K_2$$

for all $n \in N$, showing that xy is bounded.

We write m to denote the set of all bounded sequences of real numbers. Concerning this set, we have the theorem that follows.

3.4.5. Theorem

The set m is a metric space with the metric defined by

$$d\,(x, y) = \sup\ |x_n - y_n|$$

where $x = \{x_n\}$ and $y = \{y_n\}$ are in m.

Proof: If x and y are bounded, then $-y$ is bounded and $x - y = x + (-y)$ is bounded. Therefore $\sup\ |x_n - y_n|$ is a (finite) non-negative real number and hence $d\,(x, y)$ is defined and $d\,(x, y) \geq 0$.

Now $x = y \Leftrightarrow x_n = y_n$ for all $n \in N$

$\Leftrightarrow x_n - y_n = 0$ for all $n \in N$

$\Leftrightarrow |x_n - y_n| = 0$ for all $n \in N$

$\Leftrightarrow \sup\ |x_n - y_n| = 0$

$\Leftrightarrow d\,(x,y) = 0$

For any $x, y \varepsilon m$

$$d(x,y) = \sup |x_n - y_n| = \sup |-(y_n - x_n)|$$
$$= \sup |-1| \; |y_n - x_n|$$
$$= \sup |y_n - x_n| = d(y,x)$$

For any $x, y, z \varepsilon m$,

$$d(x,z) = \sup |x_n - z_n| = \sup |(x_n - y_n) + (y_n - z_n)|$$
$$\leq \sup [|x_n - y_n| + |y_n - z_n|]$$
$$\leq \sup [|x_n - y_n| + \sup |y_n - z_n|]$$
$$\leq d(x,y) + d(y,z).$$

This completes the proof of the theorem.

3.4.6. Theorem

Let $x = \{x_n\}$ and $y = \{y_n\}$ be two convergent real sequences with lim $x_n = s$ and lim $y_n = t$ and let $\alpha \varepsilon R$. Then

(i) $x + y$ is convergent and

$$\lim (x_n + y_n) = \lim x_n + \lim y_n = s + t,$$

(ii) αx is convergent and

$$\lim \alpha x_n = \alpha \lim x_n = \alpha s,$$

(iii) xy is convergent and

$$\lim (x_n y_n) = (\lim x_n)(\lim y_n) = st,$$

(iv) if $t \neq 0$ and $y_n \neq 0$ for any n, then $\frac{x}{y}$ is convergent and

$$\lim \left(\frac{x_n}{y_n}\right) = \frac{\lim x_n}{\lim y_n} = \frac{s}{t}.$$

Proof: (i) Let $\varepsilon > 0$ be given. Since $\{x_n\}$ and $\{y_n\}$ are convergent to s and t respectively, there exist positive integers n_1 and n_2 such that

$$|x_n - s| < \frac{\varepsilon}{2} \text{ for all } n \geq n_1$$

and

$$|y_n - t| < \frac{\varepsilon}{2} \text{ for all } n \geq n_2.$$

Let $n_0 = \max(n_1, n_2)$. then for all $n \geq n_0$

$$\begin{aligned}|(x_n + y_n) - (s + t)| &= |(x_n - s) + (y_n - t)|\\ &\le |x_n - s| + |y_n - t|\\ &< \frac{\varepsilon}{2} + \frac{\varepsilon}{2} = \varepsilon\end{aligned}$$

This shows that $x + y$ is convergent to $s + t$.

(ii) If $\alpha = 0$ the result clearly holds. Let $\varepsilon > 0$ be given and let $0 \neq \alpha$. Since $\{x_n\}$ is convergent to s, there exists $n_0 \in N$ such that

$$|x_n - s| < \frac{\varepsilon}{|\alpha|} \text{ for all } n > n_0.$$

Hence for $n \geq n_0$,

$$\begin{aligned}|\alpha x_n - \alpha s| &= |\alpha(x_n - s)| = |\alpha|\ |x_n - s|\\ &< |\alpha| \frac{\varepsilon}{|\alpha|} = \varepsilon\ .\end{aligned}$$

Thus αx is convergent and

$$\lim \alpha x_n = \alpha \lim x_n = \alpha s$$

(iii) Let $\{x_n\}$ and $\{y_n\}$ be convergent. Then it follows that each of them is bounded. Therefore there exist constants $K_1 > 0$ and $K_2 > 0$ such that for all $n \in N$,

$$|x_n| \leq K_1 \text{ and } |y_n| \leq K_2.$$

Now

$$\begin{aligned}|x_n y_n - st| &= |x_n (y_n - t) + t(x_n - s)|\\ &\le |x_n|\ |y_n - t| + |t|\ |x_n - s|\\ &\le K_1\ |y_n - t| + |t|\ |x_n - s|.\end{aligned}$$

Since $x_n \to s$ and $y_n \to t$ given $\varepsilon > 0$ there exist positive integers n_1 and n_2 such that

$$|x_n - s| < \frac{\frac{1}{2}\varepsilon}{|t|} \text{ for } n \geq n_1$$

and

$$|y_n - t| < \frac{\frac{1}{2}\varepsilon}{K_1} \text{ for } n \geq n_2.$$

Let $n_0 = \max(n_1, n_2)$. Then for $n \geq n_0$,

$$|x_n y_n - st| < K_1 \frac{\frac{1}{2}\varepsilon}{K_1} + |t| \frac{\frac{1}{2}\varepsilon}{|t|}$$

$$= \frac{1}{2}\varepsilon + \frac{1}{2}\varepsilon = \varepsilon.$$

This proves (iii).

(iv) Let $\lim y_n = t \neq 0$ and $y_n \neq 0$ for any n. Then there exists a positive integer n_3 such that

$$|y_n - t| < \frac{1}{2}|t| \text{ for } n \geq n_3.$$

Hence for $n \geq n_3$,

$$|t| - |y_n| \leq |y_n - t| < \frac{1}{2}|t|$$

Therefore

$$|y_n| \geq \frac{1}{2}|t| = \alpha \text{ (say) for } n \geq n_3.$$

Let $x_n \to s$. Then for $n \geq n_3$,

$$\left|\frac{x_n}{y_n} - \frac{s}{t}\right| = \left|\frac{tx_n - sy_n}{ty_n}\right|$$

$$= \left|\frac{t(x_n - x) - s(y_n - t)}{ty_n}\right|$$

$$\leq \frac{|t|\,|x_n - s| + |s|\,|y_n - t|}{|t|\,|y_n|}$$

$$\leq \frac{2}{|t|}|x_n - s| + \frac{2|s|}{|t|^2}|y_n - t|$$

Let $\varepsilon > 0$ be given. Since $x_n \to s$ and $y_n \to t$, there exist positive integers n_1 and n_2 such that

$$|x - s| < \frac{1}{4}|t|\,\varepsilon \text{ for } n \geq n_1$$

and

$$|y_n - t| < \frac{1}{4}\frac{|t|^2 \varepsilon}{|s|} \text{ for } n \geq n_2.$$

Let $n_0 = \max(n_1, n_2, n_3)$. Then for $n \geq n_0$,

$$\left|\frac{x_n}{y_n} - \frac{s}{t}\right| \leq \frac{2}{|t|}|x_n - s| + \frac{2|s|}{|t|^2}|y_n - t|$$

$$< \frac{|2|}{|t|}\frac{|t|\varepsilon}{4}\frac{2|s|}{|t|^2}\frac{|t|^2\varepsilon}{4|s|}$$

$$= \frac{\varepsilon}{2} + \frac{\varepsilon}{2} = \varepsilon$$

and this completes the proof.

We write c to denote the set of all convergent sequences of real numbers.

3.4.7. Theorem

(a) The set c is a metric space with the metric d defined by

$$d(x, y) = \sup |x_n - y_n| \text{ for } x, y \,\varepsilon\, c.$$

(b) The function ρdefined by

$$\rho(x, y) = |\lim (x_n - y_n)|$$

for $x, y \,\varepsilon\, c$ is not a metric on c.

Proof: (a) Let x and y be any two convergent sequences. Then the sequence $x-y$ is convergent and hence it is bounded. Therefore d is well defined. Now, as in Theorem 3.4.5, it follows that d is a metric on c.

(b) If x and y are convergent, $x - y$ is convergent and hence $\lim (x_n - y_n)$ exists. Therefore ρ is well defined and

$$\rho(x, y) = |\lim (x_n - y_n)| \geq 0.$$

For any $x, y, z \,\varepsilon\, c$,

$$\rho(x,x) = |\lim (x_n - x_n)| = |\lim 0| = |0| = 0$$

$$\rho(x,y) = |\lim (x_n - y_n)| = |- \lim (y_n - x_n)|$$

$$= |\lim (y_n - x_n)| = \rho(y,x)$$

$$\rho\,(x,z) = |\lim\,(x_n - z_n)\,|$$
$$= |\lim\,[(x_n - y_n) + (y_n - z_n)]\,|$$
$$= |\lim\,(x_n - y_n) + \lim\,(y_n - z_n)\,|$$
$$\leq |\lim\,(x_n - y_n)\,| + |\lim\,(y_n - z_n)\,|$$
$$= \rho\,(x,y) + \rho\,(y,z).$$

But it should be noted that ρ (x, y) = 0 does not imply that $x = y$ and hence ρ is not a metric. To prove this, let $x_n = \frac{1}{n}$ and $y_n = 0$ for all $n \varepsilon N$. The $x_n - y_n = \frac{1}{n} - 0 = \frac{1}{n}$ for all $n \varepsilon N$.

Hence

$$\lim\,(x_n - y_n) = \lim \frac{1}{n} = 0$$

and so

$$\rho\,(x, y) = |\,\lim\,(x_n - y_n)\,| = 0.$$

But $x \neq y$ and this completes the proof.

3.4.8. Theorem

If lime $x_n = s$, then

$$\lim \frac{x_1 + x_2 + \ldots + x_n}{n} = s$$

but not conversely.

Proof: To show that the converse is not true, take $x_n = (-1)^n$. Then

$$\frac{x_1 + x_2 + \ldots + x_n}{n} = \begin{cases} 0 \text{ if } n \text{ is even,} \\ -\frac{1}{n} \quad \text{if } n \text{ is odd.} \end{cases}$$

Therefore in either case,

$$\lim \frac{x_1 + x_2 + \ldots + x_n}{n} = 0$$

but the sequence $\{x_n\}$ is not convergent.

For the proof of the theroem, let $y_n = x_n - s$.

Now since $\lim x_n = s$,

$$\lim y_n = \lim(x_n - s) = \lim x_n - s$$
$$= s - s = 0.$$

Also

$$\frac{x_1 + \ldots + x_n}{n} = s + \frac{y_1 + \ldots + y_n}{n}$$

Therefore, in order to prove the theorem, it suffices to show that

$$\lim \frac{y_1 + \ldots + y_n}{n} = 0 \text{ if } \lim y_n = 0.$$

Since $\{y_n\}$ is convergent, it is bounded and hence there exists a constant $K > 0$ such that

$$|y_n| \leq K \text{ for all } n.$$

Let $\varepsilon > 0$ be given. Since $y_n \to 0$ there exists a positive integer n_1 such that

$$|y_n| < \frac{\varepsilon}{2} \text{ for } n \geq n_1.$$

Now

$$\frac{|y_1 + y_2 + \ldots + y_n|}{n} = \left|\frac{y_1 + \ldots + y_{n_1}}{n} + \frac{y_{n_1+1} + \ldots + y_n}{n}\right|$$
$$\leq \frac{|y_1| + \ldots + |y_{n_1}|}{n} + \frac{|y_{n_1+1}| + \ldots + |y_n|}{n}$$
$$\leq \frac{n_1 K}{n} + \frac{\varepsilon}{2}\frac{(n - n_1)}{n} \text{ for } n \geq n_1.$$
$$< \frac{n_1 K}{n} + \frac{\varepsilon}{2}$$

Let $n_2 \in N$ be such that $n_2 > \frac{2n_1K}{\varepsilon}$, so that $\frac{n_1 K}{n} < \frac{\varepsilon}{2}$ for $n > n_2$.

Let $n_0 = \max(n_1, n_2)$. Then for $n \geq n_0$ we have

$$\frac{|y_1 + \ldots + y_n|}{n} < \frac{\varepsilon}{2} + \frac{\varepsilon}{2} = \varepsilon,$$

and this proves the result.

3.4.9. Example

Show that

$$\lim \frac{1 + 2^{1/2} + 3^{1/3} + ... n^{1/n}}{n} = 1.$$

Let $x_n = n^{1/n}$. Then it is left to the reader to verify that

$$\lim x_n = \lim n^{1/n} = 1.$$

Therefore, by Theorem 3.4.8, we must have

$$\lim \frac{x_1 + x_2 + ... + x_n}{n} = 1$$

and thus

$$\lim \frac{1 + 2^{1/2} + 3^{1/3} + ... n^{1/n}}{n} = 1.$$

3.4.10. Definition

A sequence $\{s_n\}$ of real numbers is said to be

monotonically increasing or *nondecreasing* if

$$s_n \leq s_{n+1}, \ (n \in N)$$

and *monotonically decreasing* or *nonincreasing* if

$$s_n \geq s_{n+1}, \ (n \in N)$$

A sequence $\{s_n\}$ is said to be monotonic if either it is monotonically increasing or monotonically decreasing.

3.4.11. Theorem

A monotonic sequence of real numbers converges iff it is bounded.

Proof: We have proved in Theorem 3.1.8 that every convergent sequence in a metric space is bounded. Therefore every (monotonic) convergent sequence of real numbers is bounded.

To prove the converse, let $\{s_n\}$ be a bounded monotonically increasing sequence of real numbers. (The case of a monotonically decreasing sequence can be dealt with in a similar way). Since $\{s_n\}$ is bounded, the range of $\{s_n\}$ has a lub. Let x be the lub of the range of $\{s_n\}$. Then

$$s_n \leq x, \ (n \in N)$$

Now for every $\varepsilon > 0$ there is an integer n_0 such that

$$x - \varepsilon < s_{n_0} \leq x,$$

for otherwise $x - \varepsilon$ would be an upper bound of the range of $\{s_n\}$, which would contradict the fact that x is the lub. Since $\{s_n\}$ is monotonically increasing we have for $n \geq n_0$,

$$x - \varepsilon < s_n \leq x,$$

showing that $\{s_n\}$ converges to x, This completes the proof.

3.4.12. Example

(a) The sequence $\{s_n\}$ where $s_n = \left(1 + \frac{1}{n}\right)^n$ is bounded, monotonically increasing and hence is convergent.

$$s_n = \left(1 + \frac{1}{n}\right)^n = 1 + n \cdot \frac{1}{n} + \frac{n(n-1)}{2\,!} \frac{1}{n^2} + \ldots + \frac{n(n-1) \ldots 3.2.1.}{n\,!} \frac{1}{n^n}$$

$$= 1 + 1 + \frac{1}{2!}\left(1 - \frac{1}{n}\right) + \frac{1}{3\,!}\left(1 - \frac{1}{n}\right)\left(1 - \frac{2}{n}\right) + \ldots + \frac{1}{n!}\left(1 - \frac{1}{n}\right)\left(1 - \frac{2}{n}\right) \ldots \left(\frac{1 - n-1}{n}\right).$$

This shows that $\{s_n\}$ is monotonically increasing. Also

$$2 < s_n < 1 + 1 + \frac{1}{2\,!} + \frac{1}{3\,!} + \ldots . \frac{1}{n\,!}$$

$$< 1 + 1 + \frac{1}{2} + \frac{1}{2^2} + \ldots . + \frac{1}{2^{n-1}}$$

$$< 3.$$

Thus $\{s_n\}$ is bounded. Therefore $\{s_n\}$ converges, its limit lies between 2 and 3 and the limit is denoted by e.

(b) The sequence $\{s_n\}$ given by

$$s_n = 1 + \frac{1}{2\,!} + \frac{1}{3\,!} + \ldots + \frac{1}{n\,!}$$

is convergent.

Here $s_n - s_{n-1} = \dfrac{1}{n\,!} > 0$ for all n.

Hence $\{s_n\}$ is monotonically increasing. Further

$$\begin{aligned} s_n &= 1 + \frac{1}{2\,!} + \frac{1}{3\,!} + \ldots + \frac{1}{n\,!} \\ &< 1 + \frac{1}{2} + \frac{1}{2^2} + \ldots + \frac{1}{2^{n-1}} \\ &= \frac{1 - \frac{1}{2^n}}{1 - \frac{1}{2}} = 2\left(1 - \frac{1}{2^n}\right) \\ &= 2 - \frac{1}{2^{n-1}} < 2 \end{aligned}$$

Thus it is bounded and hence is convergent.

3.4.13. Definition

Let $\{x_n\}$ be a given sequence in a metric space (X, d) and let $\{n_k\}$ be a sequence of positive integers such that $n_1 < n_2 < \ldots$. Then the sequence $\{x_{n_k}\} = (x_{n_1}, x_{n_2}, \ldots)$ is called a *subsequence* of the given sequence $\{x_n\}$.

3.4.14. Theorem

Let (X, d) be a compact metric space and let $\{x_n\}$ be a sequence in X. Then there exists a subsequence of $\{x_n\}$ which converges to a point of X.

Proof: Let A denote the range of the sequence $\{x_n\}$. If A is a finite set, then there is a point $x \in A$ and a sequence $\{n_k\}$ of positive integers with $n_1 < n_2 < \ldots$, such that

$$x_{n_1} = x_{n_2} = \ldots = x.$$

Now the subsequence $\{x_{n_k}\}$ converges to x.

If A is infinite, since X is compact, it follows from Theorem 2.4.5 that A has a limit point $x \varepsilon X$. Therefore for every $r > 0$, $S_r(x)$ contains points of A other than x (if $x \varepsilon A$). Let $x_{n_k} \varepsilon S_{1/k}(x)$ for each $k \varepsilon N$. It now follows that the subsequence $\{x_{n_k}\}$ of $\{x_n\}$ converges to $x \varepsilon X$ and this proves the result.

3.4.15. Corollary

Every bounded real sequence contains a convergent subsequence.
Proof: Let $\{x_n\}$ be a bounded real sequence in R. Then it lies in a closed and bounded interval $[a, b] \subset R$. But by the Heine-Borel Theorem $[a,b]$ is compact. Therefore it follows from the above theorem that $\{x_n\}$ has a convergent subsequence.

3.5 LIMIT SUPERIOR AND LIMIT INFERIOR

3.5.1. Definition

Let $\{s_n\}$ be a sequence in R. Let E be the set of all x (x would be a real number or ∞ or $-\infty$) such that $s_{n_k} \to x$ for some subsequence $\{s_{n_k}\}$ of $\{s_n\}$. We define the upper limit or limit superior (lim sup in short) of $\{s_n\}$ to be sup E. Similarly, inf E is called the lower limit or limit inferior (lim inf) of $\{s_n\}$. To be precise,

$$\limsup s_n = \sup E,$$

$$\liminf s_n = \inf E,$$

3.5.2. Examples

(a) Let $s_n = (-1)^n$. Then

$$\limsup s_n = 1, \ \liminf s_n = -1.$$

(b) Let $s_n = 1 + (-1)^n$. Then

$$\limsup s_n = 2, \ \liminf s_n = 0.$$

(c) Let $s_n = \dfrac{(-1)^{n-1}}{n!}$. In this case

$$\limsup s_n = \liminf s_n = 0.$$

The sequence converges and $\lim s_n = 0$.

(d) Let $s_n = \dfrac{(-1)^n}{1+\dfrac{1}{n}}$.

Then $\lim\sup s_n = 1$, $\lim\inf s_n = -1$.

(e) Let $\{s_n\}$ be the sequence consisting of all rational numbers.

Then

$$\lim\sup s_n = \infty, \ \lim\inf s_n = -\infty.$$

It is evident from the above examples that the concepts of limit superior and limit inferior are associated with all the sequences of real numbers whether convergent or not. The two limits are, however, distinct and finite in non-convergent sequences having convergent subsequences.

Let us consider a sequence $\{s_n\}$ of real terms. If this sequence is not bounded above then we have

$\lim\sup s_n = M$, where M is large. Similarly if $\{s_n\}$ is not bounded below, we have

$$\lim\inf s_n = -M$$

Now we consider the case when the sequence $\{s_n\}$ is bounded above, which implies that the sequence is either convergent or not bounded below. In that case the sequence

$$s_n, s_{n+1}, s_{n+2}, \ldots$$

is bounded above and hence has a least upper bound μ_n, i.e.

$$\mu_n = \text{lub } \{s_n, s_{n+1}, s_{n+2}, \ldots\}$$

Moreover, it is easy to see that $\mu_n \geq \mu_{n+1}$ as

$$\{s_{n+1}, s_{n+2}, s_{n+3}, \ldots\} \subset \{s_n, s_{n+1}, s_{n+2}, \ldots\}$$

Thus the sequence $\{\mu_n\}$ is non increasing and thus either converges or diverges to some $-M$. Precisely,

(i) Convergence of $\{\mu_n\}$ implies that

$$\lim\sup s_n = \lim_{n\to\infty} \mu_n$$

(ii) Divergence of $\{\mu_n\}$ implies that

$$\limsup \mu_n = -M.$$

Further, if the sequence $\{s_n\}$ of real terms is bounded below and

$$\lambda_n = \text{glb}\,\{s_n, s_{n+1}, s_{n+2}, \ldots\}$$

then $\{s_n\}$ is nondecreasing and either the sequence is convergent so that

$$\limsup s_n = \lim_{n\to\infty} \lambda_n$$

or the sequence diverges to M.

3.5.3. Theorem

If the sequence $\{s_n\}$ of real terms converges to s then

(a) $\limsup \{s_n\} = s$

(b) $\liminf s_n = s$.

Proof: As the sequence $\{s_n\}$ converges to s, for a given $\varepsilon > 0$ there exists a positive integer n_0 such that

$$|s_n - s| < \varepsilon \text{ for } n \geq n_0$$

This means

$$s - \varepsilon < s_n < s + \varepsilon \text{ for } n \geq n_0$$

and $s + \varepsilon$ is an upper bound of

$$\{s_n, s_{n+1}, s_{n+2}, \ldots\}$$

Also, if μ_n is the least upper bound of the above set, then

$$s - \varepsilon \leq \mu_n \leq s + \varepsilon$$

Further, the sequence $\{\mu_n\}$ is nonincreasing, implying that

$$s - \varepsilon \leq \lim_{n\to\infty} \mu_n \leq s + \varepsilon$$

But

$$\lim_{n\to\infty} \mu_n = \limsup s_n,$$

Therefore,

$$s - \varepsilon \leq \limsup s_n \leq s + \varepsilon$$

Hence

$$\lim\sup s_n = s$$

This proves part (a) of the theorem. The proof of part (b) is on the same lines and is left to the reader.

3.5.4. Theorem

Let $\{s_n\}$ be a sequence of real terms, then

$$\lim\inf s_n \leq \lim\sup s_n$$

Proof: If the sequence $\{s_n\}$ is bounded, and we define

$$\lambda_n = \text{glb}\ \{s_n, s_{n+1}, s_{n+2}, \ldots\}$$

$$\mu_n = \text{lub}\ \{s_n, s_{n+1}, s_{n+2}, \ldots\}$$

then $$\lambda_n \leq \mu_n$$

Also we have

$$\lambda_n = \lim\inf s_n \text{ and } \mu_n = \lim\sup s_n$$

as the sequence $\{s_n\}$ is both bounded below and bounded above. Therefore

$$\lim\inf s_n \leq \lim\sup s_n.$$

For the case when $\{s_n\}$ is not bounded, either

$$\lim\inf s_n = -M \text{ or } \lim\sup s_n = M$$

where M is large and positive. This means either

$$-M < \lim\sup s_n \text{ or } \lim\inf s_n < M$$

Hence the theorem.

These theorems can be extended to show that a bounded sequence of real numbers has a subsequence which converges. The original sequence may or may not be convergent.

3.6 SERIES

An expression of the form

$$a_1 + a_2 + \ldots + a_n + \ldots$$

where each a_i, $i \varepsilon N$, is a real number is called an infinite series, or simply a series, of real numbers, the numbers a_i are called the terms of the series. We write $\sum_{n=1}^{\infty} a_n$ or simply Σa_n to denote the infinite series

$$a_1 + a_2 + \ldots + a_n + \ldots$$

Given an infinite series

$$\Sigma a_n$$

we associate with it a sequence $\{s_n\}$ where s_n denotes the sum of first n terms of the series Σa_n. Thus

$$s_n = a_1 + a_2 + \ldots + a_n$$

for each $n \varepsilon N$, that is,

$$s_1 = a_1$$

$$s_2 = a_1 + a_2$$

$$s_3 = a_1 + a_2 + a_3$$

and so on. This sequence $\{s_n\}$ is called the sequence of partial sums of the series Σa_n where s_n is called the nth partial sum.

3.6.1. Definition

An infinite series Σa_n is said to be convergent if the sequence $\{s_n\}$ of partial sums is convergent and in this case the limit of the sequence $\{s_n\}$ is called the sum of the series Σa_n . Otherwise the series is said to be divergent.

Example: The geometric series $\sum_{n=1}^{\infty} r^{n-1}$ is convergent for $0 \le r < 1$, for in this case $s_n = \frac{1-r^n}{1-r} \to \frac{1}{1-r}$ as $n \to \infty$.

3.6.2. Theorem

If a series Σa_n is convergent, then $\lim a_n = 0$ but the converse is not true.

Proof: Let Σa_n be convergent and let

$$s_n = a_1 + a_2 + \ldots + a_n$$

denote the sequence of its partial sums. Since Σa_n is convergent, the sequence $\{s_n\}$ converges. Therefore $\lim s_n = s$ (say). Now $a_n = s_n - s_{n-1}$. Hence

$$\begin{aligned}\lim a_n &= \lim (s_n - s_{n-1}) \\ &= \lim s_n - \lim s_{n-1} \\ &= s - s = 0.\end{aligned}$$

To show that the converse is not true, consider the series $\Sigma \frac{1}{n}$. In this case $\lim \frac{1}{n} = 0$, but the series does not converge. We shall first establish the following result and then show that the series $\Sigma \frac{1}{n}$ diverges.

3.6.3. Theorem

A series Σa_n is covergent iff for every $\varepsilon > 0$ there exists a positive integer n_0 such that

$$|a_{n+1} + \ldots + a_{n+m}| < \varepsilon \text{ for } n \geq n_0,\ m \geq 1.$$

This is known as *Cauchy's general principle of convergence.*

Proof: Let

$$s_n = a_1 + a_2 + \ldots + a_n.$$

Since Σa_n is convergent, the sequence $\{s_n\}$ is convergent and hence it is Cauchy. Therefore for every $\varepsilon > 0$ there exists n_0 such that for all $n \geq n_0$ and $m \geq 1$.

$$|s_{n+m} - s_n| = |a_{n+1} + \ldots + a_{n+m}| < \varepsilon$$

Conversely, assume that for every $\varepsilon > 0$ there exists $n_0 \in N$ such that for $n \geq n_0$, $M \geq 1$,

$$|s_{n+m} - s_n| = |a_{n+1} + \ldots + a_{n+m}| < \varepsilon$$

It now follows that $\{s_n\}$ is a Cauchy sequence of real numbers and hence is convergent. Therefore Σa_n is convergent and this completes the proof.

3.6.4. Examples

(a) The series $\Sigma \frac{1}{n}$ is divergent.

Let us assume that the series $\Sigma \frac{1}{n}$ converges. Then for every $\varepsilon > 0$ (say for $\varepsilon = 1/10$) there exists a positive integer n_0 such that for $n \geq n_0$ and $m \geq 1$,

$$\left| \frac{1}{n+1} + \dots + \frac{1}{n+m} \right| < \varepsilon.$$

Take $n = n_0$ and $m = n_0$. Then

$$\frac{1}{n+1} + \frac{1}{n+2} + \dots + \frac{1}{n+m} = \frac{1}{n_0+1} \frac{1}{n_0+2} + \dots + \frac{1}{2n_0}$$

$$> n_0 \cdot \frac{1}{2n_0} = \frac{1}{2} > \varepsilon$$

This contradiction proves the result.

(b) Show that the series

$$\frac{1}{2} + \frac{2}{3} + \frac{3}{4} + \dots$$

diverges. In this case

$$a_n = \frac{n}{n+1}$$

$$\lim a_n = \lim \frac{n}{n+1} = \lim \frac{1}{1+\frac{1}{n}} = 1 \neq 0,$$

Therefore the given series diverges, for if the series Σa_n would converge, $\lim a_n$ would be 0.

(c) $\Sigma \frac{1}{n^p}$ where $p \leq 0$ diverges, because in this case $a_n \not\to 0$ as

$n \to \infty$

Moreover $\Sigma \frac{1}{n^p}$ converges for $p > 1$ and diverges for $p \leq 1$.

We now prove that $\Sigma \frac{1}{n^p}$ converges for $p > 1$. Let

$$s_n = \frac{1}{1^p} + \frac{1}{2^p} + \frac{1}{3^p} + \ldots + \frac{1}{n^p}$$

Since $p > 1$, the sequence $\left\{\frac{1}{n^p}\right\}$ is nondecreasing. For $m \varepsilon N$ we have

$$s_{2^{m+1}-1} = 1 + \left(\frac{1}{2^p} + \frac{1}{3^p}\right) + \ldots$$

$$+ \left(\frac{1}{(2^m)^p} + \ldots + \frac{1}{(2^{m+1} - 1)^p}\right)$$

$$\leq 1 + \frac{2}{2^p} + \frac{4}{4^p} + \ldots + \frac{2^m}{(2^m)^p}$$

$$= 1 + 2^{1-p} + 4^{1-p} + \ldots + (2^m)^{1-p}$$

$$= 1 + 2^{1-p} + (2^{1-p})^2 + \ldots + (2^{1-p})^m$$

$$= \frac{1 - (2^{1-p})^{m+1}}{1 - 2^{1-p}} < \frac{1}{1 - 2^{1-p}}$$

since $2^{1-p} > 0$. For given $n \varepsilon N$, it can be shown by mathematical induction that

$$n < n + 1 \leq 2^n < 2^{n+1} - 1.$$

Also since $\{s_n\}$ is nondecreasing, it follows that for each $n \varepsilon N$

$$s_n < s_{2^{n+1}-1} < \frac{1}{1 - 2^{1-p}}.$$

Thus the sequence $\{s_n\}$ is bounded ; it is bounded above by $\frac{1}{1 - 2^{1-p}}$ and hence the series $\Sigma \frac{1}{n^p}$ is convergent.

We have proved that the series $\Sigma \frac{1}{n}$ is divergent. We can prove the same result also as follows:

Let us write the terms of the series in greater detail. Let

$$1 + \frac{1}{2} + \left(\frac{1}{3} + \frac{1}{4}\right) + \left(\frac{1}{5} + \frac{1}{6} + \frac{1}{7} + \frac{1}{8}\right) + \ldots. \quad (1)$$

Consider another series as follows

$$1+\frac{1}{2}+\left(\frac{1}{4}+\frac{1}{4}\right)+\left(\frac{1}{8}+\frac{1}{8}+\frac{1}{8}+\frac{1}{8}\right)+\ldots. \tag{2}$$

Each term of the series (1) is greater than or equal to the corresponding term of the series (2). Let s_n^1 and s_n^2 be the partial sums of the series (1) and (2) respectively. Then for $n > 2$,

$$s_n^1 > s_n^2.$$

If $n = 2^k$ for $k \varepsilon N$ we have

$$s_{2^k}^2 = 1 + k \times \frac{1}{2}$$

and hence

$$\lim_{n\to\infty} s_n^2 = \infty$$

Hence

$$\lim_{n\to\infty} s_n^1 = \infty$$

and thus $\Sigma \frac{1}{n}$ diverges.

There are several tests for the convergence of series—comparison test, D' Alembert's ratio test, Cauchy's root test, Raabe's test, logarithmic test, Cauchy's integral test, and Gauss test to mention a few. Readers are probably familiar with all these tests and have spent a considerable amount of time in exercising them. Therefore we shall discuss them here only briefly and then close this chapter with a brief discusision of absolute convergence.

3.6.5. Comparison Test

(a) Let Σu_n be a convergent series and Σx_n be another series such that $|x_n| \le ku_n$ for all $n \ge N$, where N is some integer and k is a constant. Then the series Σx_n is also convergent.

(b) If Σu_n is divergent and $|x_n| \ge ku_n$, then Σx_n diverges.

Proof: we have

$$|\sum_{j=1}^{k} u_{M+j}| < \varepsilon/k \text{ for } M \geq n_\varepsilon \text{ and} K \geq 0$$

where ε. > 0 and N_ε depends on ε: We take $M > N$.

Then

$$|\sum_{j=1}^{k} x_{M+j}| \leq \sum_{j=1}^{k} |x_{M+j}| \leq k \sum_{j=1}^{k} u_{M+j}$$

$$< \varepsilon \text{ for } M \geq N_\varepsilon \text{ and } K \geq 0.$$

Hence the series Σx_i is convergent. This proves (a).

(b) can be proved similarly.

The series Σn^{-p} is convergent for $p > 1$. This series is widely used for the comparison test.

We shall now discuss the limit form of comparison test.

Theorem: If Σu_n and Σv_n are series of positive terms such that $\lim_{n\to\infty} \frac{u_n}{v_n} = l$, l is a (finite) non-zero real number, then the two series converge or diverge together.

Proof: Clearly $l > 0$. Choose $\varepsilon > 0$ such that $l - \varepsilon > 0$. since $\lim_{n\to\infty} \frac{u_n}{v_n} = l$. there exists a positive integer m such that for all $n \geq m$,

$$\left|\frac{u_n}{v_n} - l\right| < \varepsilon$$

$$\Rightarrow l - \varepsilon < \frac{u_n}{v_n} < l + \varepsilon$$

$$\Rightarrow (l - \varepsilon)\, v_n < u_n < (l + \varepsilon)\, v_n.$$

Let Σv_n be convergent. Since $u_n < (l + \varepsilon)\, v_n$, it follows that Σu_n is convergent. Again if Σv_n is divergent, since $u_n > (l - \varepsilon)\, v_n$, Σu_n is divergent.

3.6.6. Ratio Test

Let all x_i s be positive and

$$\lim_{n \to \infty} x_{n+1}/x_n = \lambda \qquad \dots \quad (1)$$

Then $\Sigma\, x_n$ is convergent or divergent when $\lambda < 1$ or $\lambda > 1$ respectively. The test gives no information for $\lambda = 1$.

Proof: In view of (1) we have $x_{i+j} < \lambda^j x_i$

$$x_i < \alpha\, \lambda^i \text{ for } n \geq N_0 \text{ and } \alpha = x_{N_0}\, \lambda^{-N_0},$$

N_0 is some fixed number. The series $\Sigma\, \lambda^i$ is convergent for $\lambda < 1$. Hence the given series is also convergent for $\lambda < 1$.

The condition $\lambda > 1$ implies that $x_{i+1} > x_i$ for $i \geq N_0$ and $x_i \geq 0$. This means that $\lim\limits_{n \to \infty} x_i \neq 0$.

Hence the series is divergent for $\lambda > 1$.

3.6.7. Root Test

Let $\Sigma\, x_n$ be a series of positive terms and $\mu = \lim\limits_{n\to\infty} (x_n)^{\frac{1}{n}}$

Then $\mu < 1$ or $\mu > 1$ implies that the series is convergent or divergent respectively, while $\mu = 1$ does not signify anything.

Proof: From above we have

$$(x_n)^{\frac{1}{n}} < \mu + \varepsilon$$

$$\text{or} \qquad x_n < (\mu + \varepsilon)^n \text{ for } n \geq N,$$

N is an arbitrary number chosen in such a manner that

$$\mu < 1 \Rightarrow \mu + \varepsilon < 1$$

Thus the series $\Sigma\, (\mu + \varepsilon)^n$ is convergent and $x_n < (\mu + \varepsilon)^n$ for $n > N$. Therefore the series $\Sigma\, x_n$ is also convergent.

Similarly, if $\mu > 1$ we take N such that

$$(x_n)^{\frac{1}{n}} > \mu - \varepsilon \text{ or } x_n > (\mu - \varepsilon)^n \text{ for } i > N.$$

We choose ε such that $\mu - \varepsilon > 1$ and hence the series $\Sigma\, (\mu - \varepsilon)^n$ is divergent. Therefore the series $\Sigma\, x_i$ is divergent for $\mu > 1$.

3.6.8. Examples

(a) The series

$$\sum_{n=1}^{\infty} \frac{1}{n^2+1} = \frac{1}{2} + \frac{1}{5} + \frac{1}{10} + \frac{1}{17} + \ldots.$$

is convergent as the series $\sum_{n=1}^{\infty} \frac{1}{n^2}$ is convergent and

$$\frac{1}{n^2+1} < \frac{1}{n^2} \text{ for all } n = 1, 2, \ldots$$

Similarly, the series

$$\sum_{n=1}^{\infty} \frac{1}{\sqrt{n}} = 1 + \frac{1}{\sqrt{2}} + \frac{1}{\sqrt{3}} + \ldots.$$

is divergent since

$$\frac{1}{n} < \frac{1}{\sqrt{n}} \text{ for all } n = 1, 2, 3, \ldots$$

Therefore

$$\sum_{n=1}^{\infty} \frac{1}{\sqrt{n}} \text{ diverges as } \sum_{n=1}^{\infty} \frac{1}{n} \text{ diverges.}$$

Of course the divergence of $\Sigma \frac{1}{\sqrt{n}}$ follows from the general result that $\Sigma \frac{1}{n^p}$ diverges for $p \leq 1$.

(b) The series

$$\Sigma \frac{n}{2^n} = \frac{1}{2} + \frac{2}{2^2} + \frac{3}{3^3} + \ldots$$

is convergent since

$$\lim_{n\to\infty} \left(\frac{a_{n+1}}{a_n}\right) = \lim_{n\to\infty} \frac{(n+1)/2^{n+1}}{n/2^n}$$

$$\lim_{n\to\infty} \frac{1 + 1/n}{2} = \frac{1}{2} < 1.$$

(c) We consider the series

$$\sum_{n=1}^{\infty}\left(\frac{1}{n^n}\right)=1+\frac{1}{4}+\frac{1}{27}+\dots$$

Here

$$\lim_{n\to\infty}\left(\frac{1}{n^n}\right)^{1/n} = \lim_{n\to\infty}\left(\frac{1}{n}\right)=0$$

This means that $\mu < 1$. Hence the given series is convergent.

(d) Test the convergence of

$$\sum_{n=2}^{\infty}\frac{1}{(\log n)^{\rho}}=\frac{1}{(\log 2)^{\rho}}+\frac{1}{(\log 3)^{\rho}}+\dots,\ \rho>0.$$

Solution: We consider the divergent series

$$\sum_{n=1}^{\infty}\frac{1}{n}.$$

we have

$$\lim_{n\to\infty}\frac{1}{n}\bigg/\frac{1}{(\log n)^{p}}=\lim_{n\to\infty}\frac{(\log n)^{p}}{n}=0.$$

Thus $$\frac{1}{(\log n)^{p}}>\frac{1}{n},\ n>1.$$

Hence the given series is divergent.

(e) Test the convergence of $\sum_{n=1}^{\infty}\frac{1}{n!}$

we have

$$\frac{1}{2!}=\frac{1}{2},\ \frac{1}{3!}<\frac{1}{2^2},\ \frac{1}{4!}<\frac{1}{2^3},\ \dots.$$

$$\frac{1}{n!}<\frac{1}{2^{n-1}},\ \dots.$$

So $1+\frac{1}{2!}+\frac{1}{3!}+\dots+\frac{1}{n!}+\dots<1+\frac{1}{2}+\frac{1}{2^2}+\dots.$

But $1 + \frac{1}{2} + \frac{1}{2^2} + ...$ is a convergent geometric series and hence the given series is convergent.

We now discuss another test known as the Logarithmic test.

Logarithmic Test: Let $\Sigma\ u_n$ be a series of positive terms such that

$$\lim_{n\to\infty} \left(n \log \frac{u_n}{u_{n+1}}\right) = l.$$

Then the series conveges for $l > 1$ and diverges for $l < 1$. Suppose $l > 1$ Choose ε such that $l - \varepsilon > 1$.

Let $l - \varepsilon = \alpha > 1$. Since $\lim_{n\to\infty} \left(n \log \frac{u_n}{u_{n+1}}\right) = l$,

there exists a positive integer m such that for all $n > m$,

$$l - \varepsilon < n \log \frac{u_n}{u_{n+1}} < l + \varepsilon$$

$$\Rightarrow n \log \frac{u_n}{u_{n+1}} > \alpha$$

$$\Rightarrow \frac{u_n}{u_{n+1}} > e^{\frac{\alpha}{n}}.$$

since $\left\{\left(1 + \frac{1}{n}\right)^n\right\}$ is a monotonic increasing sequence converging to e,

$$\left(\frac{1+1}{n}\right)^n \leq e \quad \text{for all } n$$

and hence for all $n \geq m$,

$$\frac{u_n}{u_{n+1}} > \left(1 + \frac{1}{n}\right)^\alpha = \frac{(n+1)^\alpha}{n^\alpha} = \frac{v_n}{v_{n+1}}$$

where $v_n = \frac{1}{n^\alpha}$. But since $\Sigma\ v_n$ converges for $l > 1$, by comparison test we conclude that $\Sigma\ u_n$ is convergent. Similarly it can be shown that $\Sigma\ u_n$ diverges for $l < 1$ and this completes the proof.

Example: Test for convergence the series

$$\frac{x}{1\,!}+\frac{2^2\,x^2}{2\,!}+\frac{3^3\,x^3}{3\,!}+\ldots$$

we have

$$u_n=\frac{n^n\,x^n}{n\,!}$$

In this case

$$\lim_{n\to\infty}\frac{u_{n+1}}{u_n}=\lim_{n\to\infty}\left(\frac{n+1}{n}\right)^n x=ex$$

By the ratio test it follows that the series converges for $x<\frac{1}{e}$ and diverges for $x>\frac{1}{e}$. But for $x=\frac{1}{e}$, the ratio test fails and the logarithmic test will be useful in this case.

For $x=\frac{1}{e}$

$$\frac{u_n}{u_{n+1}}=\left(\frac{n}{n+1}\right)^n e$$

Now

$$\lim_{n\to\infty}\left(n\log\frac{u_n}{u_{n+1}}\right)=\lim_{n\to\infty} n\left[1-n\log\left(1+\frac{1}{n}\right)\right]$$

$$=\lim_{n\to\infty} n\left[1-n\left(\frac{1}{n}-\frac{1}{2n^2}+\frac{1}{3n^3}-\ldots\right)\right]$$

$$=\frac{1}{2}<1.$$

Hence by the logarithmic test the series is shown to diverge for $x=\frac{1}{e}$.

3.6.9. Definition

An infinite series $\Sigma\, a_n$ with real terms is said to be absolutely convergent if the series $\Sigma\, |a_n|$ is convergent. We shall observe in the following theorem that every absolutely convergent series is convergent but the converse is not true. This means that there are series which are convergent but not absolutely convergent. Such series are called *conditionally convergent.*

3.6.10. Theorem

Every absolutely convergent series of real terms is convergent, but the converse is not true.

Proof: Let $\Sigma\, a_n$ be absolutely convergent, so that $\Sigma\, |a_n|$ is convergent. Therefore, for a given $\varepsilon > 0$ there exists $n_0\ \varepsilon\ N$ such that for $n \geq n_0$ and $m \geq 1$,

$$|a_{n+1}| + |a_{n+2}| + \ldots + |a_{n+m}| < \varepsilon.$$

Now

$$|a_{n+1} + a_{n+2} + \ldots + a_{n+m}| < \varepsilon$$

for $n \geq n_o$ and $m \geq 1$. Therefore the series $\Sigma\, a_n$ converges.

Before proving that the converse is not true we shall first establish a theorem which gives the test for convegence of an alternating series

$$u_1 - u_2 + u_3 - u_4 + \ldots$$

where each $u_n > 0$.

3.6.11. Leibniz Theorem

If for the series

$$u_1 - u_2 + u_3 - u_4 + \ldots \qquad (u_n > 0)$$

the terms are such that

$$u_1 > u_2 > u_3 > \ldots$$

$$\lim_{n \to \infty} u_n = 0,$$

then the series converges, its sum is positive and does not exceed the first term.

Proof: Let

$$S_{2m} = (u_1 - u_2) + (u_3 - u_4) + \ldots + (u_{2m-1} - u_{2m})$$

Then it follows that

$$S_{2m} > 0.$$

Also

$$S_{2m} = u_1 - (u_2 - u_3) - (u_4 - u_5) - \ldots - (u_{2m-2} - u_{2m-1}) - u_{2m}$$

From this it now follows that

$$S_{2m} < u_1.$$

Thus S_{2m} is an increasing function of m and is bounded above by u_1. Therefore

$$\lim_{m\to\infty} S_{2m} = s,\ 0 < s < u_1.$$

Now $S_{2m+1} = S_{2m} + u_{2m+1}$. But $\lim_{m\to\infty} u_{2m+1} = 0$. Hence

$$\lim_{m\to\infty} S_{2m+1} = \lim_{m\to\infty} (S_{2m} + u_{2m+1}) = \lim_{m\to\infty} S_{2m} + \lim_{m\to\infty} u_{2m+1}$$

$$= s + 0 = s.$$

This completes the proof.

To show that there are series which are convergent, but not absolutely convergent, take $a_n = \frac{(-1)^n}{n}$, Then $\Sigma\, a_n$ is the series

$$1 - \frac{1}{2} + \frac{1}{3} - \frac{1}{4} + \ldots.,$$

which is convergent. The convergence of this series follows from Leibnitz's theorem and we leave it to the reader to establish this as an exercise. Observe that the series

$$\Sigma\, |a_n| = 1 + \frac{1}{2} + \frac{1}{3} + \ldots$$

is divergent. This example also shows that the divergence of $\Sigma\, |a_n|$ does not imply divergence of $\Sigma\, a_n$.

We have just discussed only one test (Leibnitz Theorem) for the convergence of an alternating series. We shall close this chapter with describing one important test, Abel's test, for the convergence of series with arbitratry terms which are not necessarily absolutely convergent.

For this purpose we must first prove a lemma.

Lemma: Let $\{b_n\}$ be a positive, monotonic decreasing sequence and $\{A_n\}$ a bounded sequence. Then the series $\Sigma\, A_n (b_n - b_{n+1})$ is absolutely convergent.

Proof: Since $\{A_n\}$ is bounded, there exists a positive integer K such that

$$|A_n| < K \text{ for all } n.$$

Hence

$$\sum_{n=1}^{m} | A_n (b_n - b_{n+1}) | < K \sum_{n=1}^{m} | b_n - b_{n+1} |$$

$$= K \sum_{n=1}^{m} (b_n - b_{n+1}) , \text{ (since } b_n - b_{n+1} \geq 0)$$

$$= K (b_1 - b_{m+1}) < K b_1.$$

Now $\Sigma A_n \mid (b_n - b_{n+1}) \mid$ is a series of positive terms and its sequence of partial sums is bounded above by kb_1. Hence $\Sigma \mid A_n (b_n - b_{n+1}) \mid$ is convergent and thus $\Sigma A_n (b_n - b_{n+1})$ is absoltely convergent.

Able's test: Let $\{b_n\}$ be a positive monotonic decreasing sequence and Σu_n a convergent series. Then $\Sigma u_n b_n$ is convergent.

Proof: Let $v_n = u_n b_n$, $s_n \sum_{r=1}^{n} u_r$, $v_n = \sum_{r=1}^{n} v_r$..

Now

$$v_n = u_1 b_1 + u_2 b_2 + \ldots + u_n b_n$$

$$= s_1 b_1 + (s_2 - s_1) b_2 + \ldots. + (s_n - s_{n-1}) b_n$$

$$= s_1(b_1 - b_2) + s_2 (b_2 - b_3) + \ldots + s_{n-1} (b_{n-1} - b_n) + s_n b_n$$

$$= \sum_{r=1}^{n-1} s_r (b_r - b_{r+1}) + s_n b_n.$$

since Σu_n is convergent, the sequence $\{s_n\}$ is convergent and hence bounded. Also $\{b_n\}$ is positive and monotonic decreasing. Therefore, by the above lemma, $\Sigma s_n (b_n - b_{n+1})$ is convergent. Hence the sequence of partial sums

$$\sum_{r=1}^{n-1} s_r (b_r - b_{r+1}) \to \text{a finite limit as } n \to \infty.$$

Further since $\{b_n\}$ is monotonic decreasing and bounded below by zero, $\{b_n\}$ is convergent and hence $b_n \to$ a finite limit as $n \to \infty$. Therefore $s_n b_n \to$ a finite limit as $n \to \infty$. Thus it follows that $v_n \to$ a finite limit as $n \to \infty$. Hence the series $\Sigma v_n = \Sigma u_n b_n$ converges and this completes the proof.

EXERCISES

1. Answer "Yes or "No" :

(i) The range of the sequence
(0, 1, 0, 1, 0, 1, ...)
is (0, 1)

(ii) The sequence $\left\{\frac{1}{n}\right\}$ is bounded.

(iii) Every bounded sequence is convergent.

(iv) The sequence
$\left(1, \frac{1}{2}, 1, \frac{1}{2}, \frac{1}{3}, 1, \frac{1}{2}, \frac{1}{3}, \frac{1}{4}, \ldots\right)$ is convergent.

(v) The sequence in (iv) is a Cauchy sequence.

(vi) Every Cauchy sequence in a metric space is convergent.

(vii) The set R is a complete metric space.

(viii) The set of all the irrational numbers is a metric space.

2. State which of the following are correct:

(i) $\lim\limits_{n\to\infty} s_n = s \Rightarrow \lim\limits_{n\to\infty} \frac{1}{s_n} = \frac{1}{s}$

(ii) $\lim\limits_{n\to\infty} s_n = s \Rightarrow \lim\limits_{n\to\infty} |s_n| = |s|$

(iii) $\lim\limits_{n\to\infty} \frac{1}{n^\alpha} = 0 \Rightarrow \alpha \geq 0.$

3. Correct (if necessary) the following statements:

(i) If $\{x_n\}$ and $\{y_n\}$ are bounded and real then $\left\{\frac{x_n}{y_n}\right\}$, $y_n \neq 0$, is bounded.

(ii) If $\{x_n\}$ is convergent then $\{x_n^2\}$ is also convergent.

(iii) If $\{x_n\}$ is convergent then $\left\{\frac{1}{x_n}\right\}$ is also cinvergent.

(iv) $\left\{\frac{n}{n^2+1}\right\}$ is monotonic and bounded.

(v) $\lim_{n\to\infty} a_n = 0$ implies that $\Sigma\, a_n$ is convergent.

(vi) $\Sigma \frac{1}{n}$ is divergent.

(vii) $\Sigma \frac{1}{n^p}$ is convergent, $\rho < 1$.

(viii) lim sup sin $(n\pi/2) = 1$.

4. Answer true or false:

(i) $\lim_{i\to\infty} x_i = 0 \Rightarrow \Sigma\, x_i$ is convergent.

(ii) If $\Sigma\, x_i$ is convergent and $\Sigma\, y_i$ is divergent then $\Sigma\,(x_i + y_i)$ may be convergent or divergent.

(iii) If $\Sigma\, x_i$ is convergent then $\Sigma\, |x_i|$ is also convergent.

(iv) $\Sigma\,(i)^0$ is a convergent series.

5. Test for the convergence of the following series

(i) $\frac{\alpha}{\beta} + \frac{1+\alpha}{1+\beta} + \frac{(1+\alpha)(2+\alpha)}{(1+\beta)(2+\beta)} + \ldots$

(ii) $\sum_{i=1}^{\infty} \frac{\alpha+i}{\beta+\iota}$

(iii) $\sum_{i=1}^{\infty} \frac{4i+1}{3i^3 - i^2 - 1}$

(iv) $\sum_{i=1}^{\infty} \frac{i}{2^i}$

(v) $\sum_{i=1}^{\infty} \frac{\cos(i\pi)}{i^2}$

6. Two Cauchy sequences $\{x_n\}$ and $\{y_n\}$ in a metric space (X, d) are said to be equivalent if

$$\lim_{n\to\infty} d(x_n, y_n) = 0.$$

Show that this defines an equivalence relation in the set of all Cauchy sequences in (X, d).

7. If $\{x_n\}$ and $\{y_n\}$ are two Cauchy sequences in a metric space (X, d), then show that the sequence $\{d(x_n, y_n)\}$ is convergent.
8. If $\{x_n\}$ is a sequence in a metric space (X, d) such that the series $\Sigma\, d(x_n, x_{n+1})$ converges, then prove that $\{x_n\}$ is Cauchy.
9. If $\{x_n\}$ and $\{y_n\}$ are two convergent sequences of real numbers and if

$$u_n = \max(x_n, y_n),\ v_n = \min(x_n, y_n),$$

then show that the sequences $\{u_n\}$ and $\{v_n\}$ are convergent and that

$$\lim u_n = \max(\lim x_n, \lim y_n)$$

$$\lim v_n = \min(\lim x_n, \lim y_n).$$

10. If $x = \{x_n\}$ and $y = \{y_n\}$ are two sequences of real numbers such that x is convergent and y is divergent, then show that

    (i) the sequence $x + y$ diverges,

    (ii) $\lim \dfrac{x_n}{y_n} = 0$

11. Show that a monotonic increasing (decreasing) sequence which is not bounded diverges to ∞ $(-\infty)$.
12. Prove that the convergence of a real sequence $\{x_n\}$ implies the convergence of $\{|x_n|\}$. Is the converse true? Justify by an example.
13. Show that the sequence $\{n^{1/n}\}$ converges and that

$$\lim_{n\to\infty} n^{1/n} = 1.$$

14. Show that for any real number x,

$$\lim_{n\to\infty} \frac{x^n}{n!} = 0$$

15. Show that the sequence $\{x_n\}$ where

$$x_n = 1 + \frac{1}{1!}\ \frac{1}{2!} + \ldots + \frac{1}{(n-1)!}$$

converges.

16. Show that $\{x_n\}$ where

$$x_n = 1 + \frac{1}{2^2} \; \frac{1}{3^2} + \ldots + \frac{1}{n^2}$$

is convergent and $\lim x_n \leq 2$.

[*Hint*: $\{x_n\}$ is monotonically increasing and bounded}

17. Show that $\{x_n\}$ where

$$x_n = 1 + \frac{1}{2} + \ldots + \frac{1}{n}$$

is not convergent.

[*Hint*: $\{x_n\}$ is monotonic but not bounded and hence not convergent]

18. Show that the series $\Sigma \dfrac{1}{n^p}$ converges for $p > 1$ and diverges for $p \leq 1$.

19. If the limit superior of the sequence $\{s_n\}$ is s then prove that the limit superior of any subsequence of $\{s_n\}$ is less than or equal to s.

20. If $\{s_n\}$ is bounded and $\lim\inf s_n = s$, then prove that $\{s_n\}$ has a subsequence which converges to s.

4 Continuity

4.1 LIMITS

4.1.1. Definition

Let (X, d_1) and (Y, d_2) be metric spaces and let $E \subset X$. Let f be a mapping from E into Y and let a be a limit point of E. We say that $f(x)$ tends to a limit $b \in Y$ as x tends to a if for every $\varepsilon > 0$ there exists a $\delta > 0$ such that

$$d_2(f(x), b) < \varepsilon \text{ for all } x \in E \text{ with } d_1(x, a) < \delta.$$

If $f(x)$ tends to b as x tends to a, then we denote this symbolically by

$$f(x) \to b \text{ as } x \to a$$

or by

$$\lim_{x \to a} f(x) = b.$$

Note that the point a is a limit point of E (domain of f) and belongs to X, but it need not be in E, even if $a \in E$ it may not be true that

$$\lim_{x \to a} f(x) = f(a).$$

Let $X = Y = R$, $E \subset R$ and $f: E \to R$, in other words, let f be a real function defined on a subset E of R. Let a be a limit point of E. Then $f(x) \to b \in R$ as $x \to a$ if for every $\varepsilon > 0$ there exists a $\delta > 0$ such that

$$|f(x) - b| < \varepsilon \text{ for } |x - a| < \delta \text{ and } x \in E$$

$$\text{i.e., for } a - \delta < x < a + \delta \text{ and } x \in E.$$

This motivates the following definitions for real functions with an interval domain.

A *real function* f has a left-handed limit b or tends to a limit b as x tends to a from the left if for every $\varepsilon > 0$ there exists a $\delta > 0$ such that

$$| f(x) - b | < \varepsilon \text{ for } a-\delta < x < a,$$

symbolically this is written as

$$\lim_{x \to a-} f(x) = b \text{ or } f(x) \to b \text{ as } x \to a-.$$

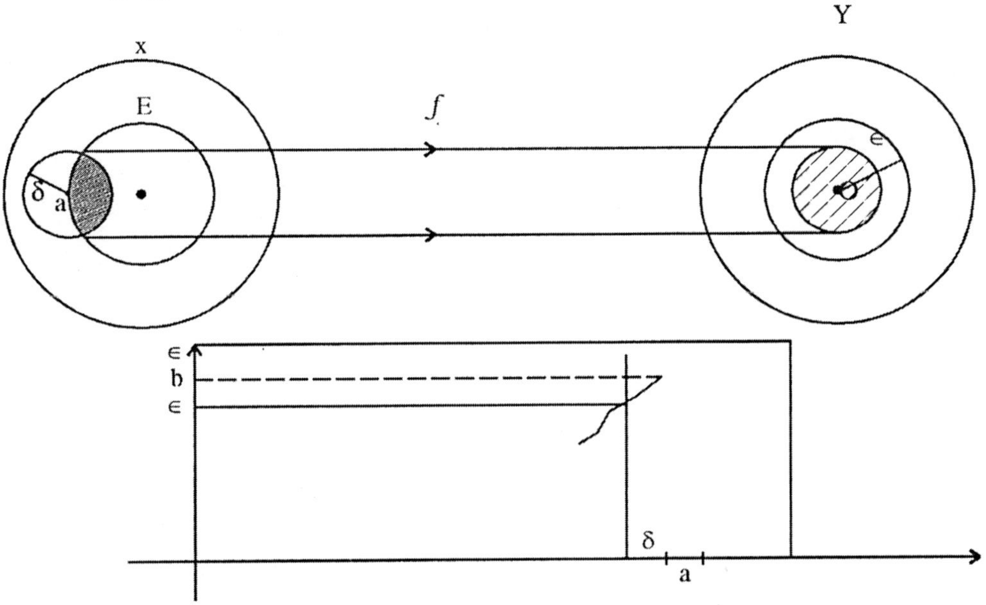

Figure 4.1

Similarly f has a right-handed limit b or tends to a limit b as x tends to a from the right if for every $\varepsilon > 0$ there exists a $\delta > 0$ such that

$$| f(x) - b | < \varepsilon \text{ for } a < x < a + \delta,$$

and symbolically this is expressed as

$$\lim_{x \to a+} f(x) = b \text{ or } f(x) \to b \text{ as } x \to a+$$

It should be noted that if both the left-handed and right-handed limits of a function at a point exist and are equal, then it is just the limit of the function at that point.

Let $X = R \times R = R^2$, $Y = R$, $E \subset R^2$ and $f : R^2 \to R$

Here f is a function of two variables, with the value $f(x, y)$ where $x \in R$ and $y \in R$. A variable point in R^2 is denoted by (x, y). Also, let $(x, y) = (a, b)$ be a limit point of domain E of f. Then the function $f(x, y)$ is said to have a simulataneous limit L, as (x, y) tends to (a, b) if for every $\varepsilon > 0$ there exists a positive number δ, such that

$$| f(x, y) - L | < \varepsilon \text{ for } (x, y) \in N_\varepsilon\ (a, b)$$

Here

$$N_\delta\ (a, b) = \{(x, y) \mid (x - a)^2 + (y - b)^2 < \delta^2\} \cap E$$

We also write

$$\lim_{(x,y) \to (a,b)} f(x,y) = L$$

It may however be noted that, if

$$\lim_{(x,y) \to (a,b)} f(x,y) = L, \ \lim_{x \to a} f(x, b) = f_a$$

and

$$\lim_{y \to b} f(a, y) = f_b$$

then it is necessary that

$$f_a = f_b = L$$

4.1.2. Examples

(a) Prove that $\lim\limits_{x \to 0} x \sin \dfrac{1}{x} = 0$

We have

$$|x \sin \frac{1}{x}| \le |x| \ |\sin \frac{1}{x}| < |x|$$

Given $\varepsilon > 0$ choose $\delta = \varepsilon$. Then it follows that

$$|x \sin \frac{1}{x}| < \varepsilon \text{ whenever } 0 < |x| < \delta.$$

So

$$\lim_{x \to 0} x \sin \frac{1}{x} = 0.$$

(b) Prove that $\lim\limits_{x \to 1} \dfrac{x^2 - 1}{x - 1} = 2.$

We have $\left|\dfrac{x^2 - 1}{x - 1} - 2\right| = \left|\dfrac{x^2 - 2x + 1}{x - 1}\right| = \left|\dfrac{(x - 1)^2}{x - 1}\right| = |x - 1|$

Given $\varepsilon > 0$ choose $\delta = \varepsilon$. Then whenever $|x - 1| < \delta$ we have

$$\left|\frac{x^2 - 1}{x - 1} - 2\right| < \varepsilon$$

and hence

$$\lim_{x \to 1} \frac{x^2 - 1}{x - 1} = 2.$$

(c) Let $f: R \to R$. If $\lim_{x \to a} f(x)$ exists, then show that it must be unique.

Assume, to the contrary, that $\lim_{x \to a} f(x) = b_1$ and $\lim_{x \to a} f(x) = b_2$. Let $\varepsilon > 0$ be given. Then we can choose $\delta_1 > 0$ and $\delta_2 > 0$ such that

$$|f(x) - b_1| < \frac{\varepsilon}{2} \text{ for } 0 < |x - a| < \delta_1$$

$$|f(x) - b_2| < \frac{\varepsilon}{2} \text{ for } 0 < |x - a| < \delta_2.$$

Let $\delta = \min(\delta_1, \delta_2)$. Then for $0 < |x - a| < \delta$ we have

$$\begin{aligned} |b_1 - b_2| &= |b_1 - f(x) + f(x) - b_2| \\ &\le |f(x) - b_1| + |f(x) - b_2| \\ &< \frac{\varepsilon}{2} + \frac{\varepsilon}{2} = \varepsilon \end{aligned}$$

Since $\varepsilon > 0$ is arbitrary we must have $b_1 = b_2$.

(d) Find the right-handed and left-handed limits of

$$f(x) = \frac{x^2 - 1}{x - 1}$$

as $x \to 1$.

For the left-handed limit put $x = 1 - h$ where $h > 0$. Hence, when $x \to 1-$, $h \to 0+$, so

$$\begin{aligned} \lim_{x \to 1-} \frac{x^2 - 1}{x - 1} &= \lim_{h \to 0+} \frac{(1 - h)^2 - 1}{-h} \\ &= \lim_{h \to 0+} \frac{-h(2 - h)}{-h} \\ &= \lim_{h \to 0+} (2 - h) = \lim_{h \to 0+} 2 - \lim_{h \to 0+} h \\ &= 2 - 0 = 2. \end{aligned}$$

For the right-handed limit, put $x = 1 + h$ where $h > 0$. Then $h \to 0+$ when $x \to 1+$. Now

$$\lim_{x \to 1+} \frac{x^2 - 1}{x - 1} = \lim_{h \to 0+} \frac{(1+h)^2 - 1}{h}$$

$$= \lim_{h \to 0+} (2 + h) = 2.$$

Thus both right-handed and left-handed limits exist and are equal. Therefore

$$\lim_{x \to 1} \frac{x^2 - 1}{x - 1} \text{ exists and is equal to } 2$$

In fact we can calculate the limit also in the following manner:

$$\lim_{x \to 1} \frac{x^2 - 1}{x - 1} = \lim_{x \to 1} \frac{(x+1)(x-1)}{x - 1}$$

$$= \lim_{x \to 1} (x + 1) \text{ (since } x \neq 1)$$

$$= 2.$$

4.1.3. Theorem

Let (X, d_1) and (Y, d_2) be metric spaces, $E \subset X$ and $f : E \to Y$. Let a be a limit point of E and $b \in Y$. Then,

$$\lim_{x \to a} f(x) = b$$

if

$$\lim_{n \to \infty} f(a_n) = b$$

for every sequence $\{a_n\}$ in E, $a_n \neq a$ for any n, converging to a.

Proof: Suppose that $\lim_{x \to a} f(x) = b$. Let $\varepsilon > 0$ be given. Then there exists a $\delta > 0$ such that

$d_2(f(x), b) < \varepsilon$ with $d_1(x, a) < \delta$. Choose a sequence $\{a_n\}$ in E such that $a_n \neq a$ for any $n \in N$ and $\lim_{n \to \infty} a_n = a$. Then there exists $n_0 \in N$ such that

$$d_1(a_n, a) < \delta \text{ for } n > n_0.$$

Hence for $n > n_0$ we have

$$d_2(f(a_n), b) < \varepsilon$$

and this means that $\lim_{n\to\infty} f(a_n) = b$.

Conversely, suppose that $\lim_{x\to a} f(x) \neq b$. Then there exists some $\varepsilon > 0$ such that for every $\delta > 0$ there exists a point $x \,\varepsilon\, E$ (depending on δ) with the property that

$$d_2\,(f(x),\, b) \geq \varepsilon \text{ but } d_1\,(x, a) < \delta.$$

Now taking $\delta_n = \frac{1}{n}$ we can find a sequence $\{a_n\}$ in E such that $a_n \neq a$ for any n and $\lim_{n\to\infty} a_n = a$ but $\lim_{n\to\infty} f(a_n) \neq b$.

This completes the proof of the theorem.

4.1.4. Definition

Let f and g be any two real functions defined on a subset A of R and let $\alpha \,\varepsilon\, R$. Then we define new functions $f + g$, $f - g$, αf, fg and $\frac{f}{g}$ on A by

$$(f+g)\,(x) = f(x) + g\,(x),$$
$$(f-g)\,(x) = f(x) - g\,(x),$$
$$(\alpha f)\,(x) = \alpha\, f(x),$$
$$(fg)\,(x) = f(x)\; g\,(x), \quad \text{and}$$
$$\left(\frac{f}{g}\right)(x) = \frac{f(x)}{g\,(x)} \text{ if } g\,(x) \neq 0 \text{ for any } x \,\varepsilon\, A..$$

4.1.5. Theorem

Let f and g be real functions defined on $A \subset R$ and let $c \,\varepsilon\, A$ be such that

$$\lim_{x\to c} f(x) = l \text{ and } \lim_{x\to c} g\,(x) = m.$$

Then

(i) $\lim_{x\to c} (f+g)\,(x) = \lim_{x\to c} f(x) + \lim_{x\to c} g\,(x) = l + m,$

(ii) $\lim_{x\to c} (f-g)\,(x) = \lim_{x\to c} f(x) - \lim_{x\to c} g\,(x) = l - m,$

(iii) $\lim_{x\to c} (\alpha f)\,(x) = \alpha \lim_{x\to c} f(x) = \alpha\, l,\ \alpha \,\varepsilon\, R,$

(iv) $\lim_{x \to c} (fg)(x) = \lim_{x \to c} f(x) \lim_{x \to c} g(x) = lm,$

(v) $\lim_{x \to c} \left(\frac{f}{g}\right)(x) = \frac{\lim_{x \to c} f(x)}{\lim_{x \to c} g(x)} = \frac{l}{m}$ if

$\lim_{x \to c} g(x) = m \neq 0.$

Proof: The proof is similar to that of Theorem 3.4.6 and we leave it as an exercise for the reader.

4.2 CONTINUITY

4.2.1. Definition

Let (X, d_1) and (Y, d_2) be metric spaces and let f be a mapping from X into Y. f is said to be continuous at a point $a \in X$ if for every $\varepsilon > 0$ there exists a $\delta > 0$ such that

$$d_2(f(x), f(a)) < \varepsilon \text{ for } d_1(x, a) < \delta.$$

This means that for every $\varepsilon > 0$ there exists a $\delta > 0$ such that $f(x)$ belongs to the open ball in Y of radius ε and centre $f(a)$ whenever x belongs to the open ball in X of radius δ and centre a or, equivalently, that the image under f of an open ball of radius δ and centre a is contained in the open ball of radius ε and centre $f(a)$; and f is said to be continuous on X if it is *continuous* at every point of X.

If $X = Y = R$, then $f: R \to R$ is said to be continuous at $a \in R$ if for every $\varepsilon > 0$ there exists $\delta > 0$ such that

$$|f(x) - f(a)| < \varepsilon \text{ for } |x - a| < \delta.$$

In other words, f is continuous at a if for every $\varepsilon > 0$ there exists a $\delta > 0$ such that $f(x)$ belongs to the open interval $(f(a) - \varepsilon, f(a) + \varepsilon)$ whenever x belongs to the open interval $(a - \delta, a + \delta)$.

It should be noted that 'f is continuous at a point a' means that $f(x) \to$ a limit as $x \to a$ and the limit is equal to $f(a)$, i.e.,

$$\lim_{x \to a} f(x) = f(a).$$

The definition of continuity for real functions can be artificially broken up into two parts. Recall that f is continuous at a if for every $\varepsilon > 0$ there exists $\delta > 0$ such that

$$|x-a| < \delta \Rightarrow |f(x) - f(a)| < \varepsilon$$
$$\Rightarrow f(a) - \varepsilon < f(x) < f(a) + \varepsilon.$$

This motivates the following concepts:

f is said to be *lower semi-continuous* at a point a if for every $\varepsilon > 0$ exists a $\delta > 0$ such that

$$f(x) > f(a) - \varepsilon \text{ for } |x-a| < \delta;$$

f is said to be *upper semi-continuous* at a point a if for every $\varepsilon > 0$ there exists a $\delta > 0$ such that

$$f(x) < f(a) + \varepsilon \text{ for } |x-a| < \delta.$$

A function is said to be *discontinuous* at a point of its domain if it is not continuous there and such a point is called a *point of discontinuity* for the function.

4.2.2. Theorem

Let (X, d_1) and (Y, d_2) be metric spaces and let f be a mapping from X into Y. Then f is continuous at a point a ε X iff

$$\lim_{n\to\infty} f(a_n) = f(a)$$

for every sequence $\{a_n\}$ in X converging to a.

Proof: Let f be continuous at a ε X and let $\varepsilon > 0$ be given. Then there exists a $\delta > 0$ such that

$$d_2(f(x), f(a)) < \varepsilon \text{ for } 0 < d_1(x, a) < \delta.$$

Let $\{a_n\}$ be a sequence in X which converges to a. Then there exists n_0 ε N such that

$$d_1(a_n, a) < \delta \text{ for } n > n_0.$$

Hence

$$d_2(f(a_n), f(a)) < \varepsilon \text{ for } n > n_0,$$

which shows that $\lim_{n\to\infty} f(a_n) = f(a)$.

Conversely, suppose that f is not continuous at a. We shall show that there exists a sequence $\{a_n\}$ which converges to a but the sequence $\{f(a_n)\}$ does not converge to $f(a)$. Since f is not continuous at a, there exists an $\varepsilon > 0$ such that for every $\delta > 0$, there is an x with

$$d_2(f(x), f(a)) \geq \varepsilon \text{ for } d_1(x, a) < \delta$$

By taking $\delta = \frac{1}{n}$ we can get a sequence $\{a_n\}$ such that

$$d_2\,(f(a_n), f(a)) \geq \varepsilon \text{ for } d_1\,(a_n, a) < \frac{1}{n}.$$

Thus, although $\{a_n\}$ converges to a, the sequence $\{f(a_n)\}$ does not converge to $f(a)$. This completes the proof of the theorem.

4.2.3. Theorem

Let f and g be two real functions defined on a subset A of R and let f be continuous at a point $a \,\varepsilon\, A$. Then the functions $f + g$, $f - g$, αf where $\alpha \,\varepsilon\, R$, fg and f/g if $g(a) \neq 0$ are all continuous at $a \,\varepsilon\, A$.

Proof: This is similar to Theorems 3.4.6 and 4.1.5 and is left to the reader as an exercise.

4.2.4. Theorem

Let (X, d_1) and (Y, d_2) be metric spaces. A function $f : X \to Y$ is continuous iff for every open set $V \subset Y$, $f^{-1}(V)$ is open in X.

Proof: Let $f : X \to Y$ be continuous and let V be an open set in Y. We have to show that $f^{-1}(V)$ is open in X. Let $x \,\varepsilon\, f^{-1}(V)$. Then $f(x) \,\varepsilon\, V$. Since V is open and $f(x) \,\varepsilon\, V$, there exists some $\varepsilon > 0$ such that

$$S_{\varepsilon}^{d_2}(f(x)) \subset V$$

where $S_{\varepsilon}^{d_2}(f(x))$ denotes an open sphere in (Y, d_2) with centre $f(x)$ and radius ε. Since f is continuous on X, it is continuous at x and hence there exists a $\delta > 0$ such that,

$$f(S_{\delta}^{d_1}(x)) \subset S_{\varepsilon}^{d_2}(f(x))$$

where $S_{\delta}^{d_1}(x)$ denotes an open sphere in (X, d_1) with centre x and radius δ. Therefore it follows that

$$f(S_{\delta}^{d_1}(x)) \subset V$$

and so

$$S_{\delta}^{d_1}(x) \subset f^{-1}(V).$$

This shows that x is an interior point of $f^{-1}(V)$ and hence $f^{-1}(V)$ is open.

Conversely, assume that for every open set $V \subset Y$, $f^{-1}(V)$ is open in X. Let $\varepsilon > 0$ be given and $x \,\varepsilon\, X$.

Then $S_{\varepsilon}^{d_2}(f(x))$, being an open sphere, is an open set in Y. Therefore $f^{-1} S_{\varepsilon}^{d_2}(f(x))$ is an open set in X containing x, hence x is an interior point of $f^{-1} S_{\varepsilon}^{d_2}(f(x))$. Therefore there exists a $\delta > 0$ such that

$$S_{\delta}^{d_1}(x) \subset f^{-1}(S_{\varepsilon}^{d_2}(f(x))),$$

so that

$$f\, S_{\delta}^{d_1}(x) \subset S_{\varepsilon}^{d_2}(f(x)).$$

This shows that f is continuous at x. Since x is arbitrary f is continuous on X and this completes the proof.

We shall now introduce the concept of composition of two mappings and show that the composition of two continuous maps is a continuous map.

4.2.5. Definition

Let (X, d_1), (Y, d_2) and (Z, d_3) be metric spaces. Let $A \subset X$, $f: A \subset X \to Y$ and $g: f(A) \subset Y \to Z$. (Figure 4.2).

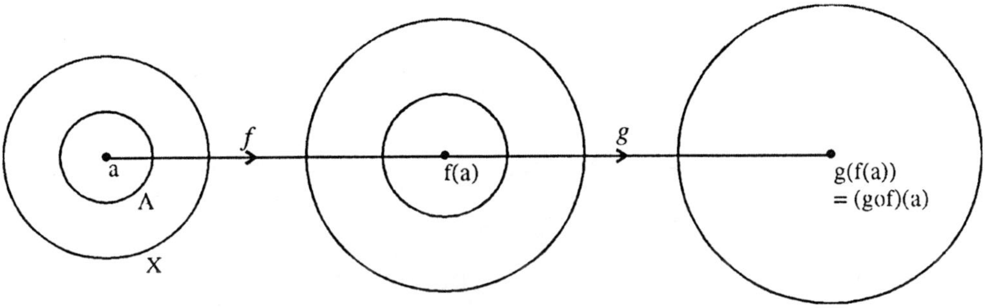

Figure 4.2

Then the composition of f and g, denoted by gof, is a maping from A into Z and is defined by

$$(gof)(a) = g(f(a)) \text{ for all } a \in A.$$

It should be noted that the composition of two mappings f and g is defined only when the range of $f \subseteq$ the domain of g.

4.2.6. Theorem

Under the notation in the above definition, if f is continuous at $a \in A$ and g is continuous at $f(a)$, then gof is continuous at $a \in A$.

Proof: Let $\varepsilon > 0$ be given. Since g is continuous at $f(a)$, there exists a $\delta_1 > 0$ such that

$$d_3(g(y), g(f(a))) < \varepsilon \text{ for } d_2(y, f(a)) < \delta, \; y \in f(A).$$

Since f is continuous at a, there exists $\delta > 0$ such that

$$d_2(f(x), f(a)) < \delta_1 \text{ for } d_1(x, a) < \delta, \; x \in A.$$

Therefore for $x \in A$, $f(x) \in f(A)$ and hence

$$d_3(g(f(x)), g(f(a))) < \varepsilon \text{ for } d_1(x, a) < \delta.$$

Hence gof is continuous at a and this proves the result.

4.2.7. Examples

(a) Show that $f : R \to R$ defined by

$$f(x) = \begin{cases} x \sin \dfrac{1}{x} & (x \neq 0), \\ 0 & (x = 0), \end{cases}$$

is continuous at $x = 0$.

In this case $f(0) = 0$. Also

$$\lim_{x \to 0} f(x) = \lim_{x \to 0} (x \sin 1/x) = 0$$

so that

$$\lim_{x \to 0} f(x) = 0 = f(0)$$

and hence f is continuous at $x = 0$.

(b) Discuss the continuity of the function $f : R \to R$ defined by

$$f(x) = \begin{cases} \dfrac{x - |x|}{x} & (x \neq 0), \\ 2 & (x = 0). \end{cases}$$

If $x < 0$, then $|x| = -x$, so that

$$f(x) = \frac{x + x}{x} = \frac{2x}{x} = 2$$

and the function is continuous for all $x < 0$. If $x > 0$, $|x| = x$, so that

$$f(x) = \frac{x - x}{x} = \frac{0}{x} = 0.$$

Hence the function is continuous for all $x > 0$. Now $f(0) = 2$. Further

$$\lim_{x \to 0-} f(x) = \lim_{x \to 0+} \frac{x + x}{x} = 2$$

and

$$\lim_{x \to 0+} f(x) = \lim_{x \to 0+} \frac{0}{x} = 0.$$

Hence f is discontinuous at $x = 0$.

(c) For $x \geq 0$, let $[x]$ denote the largest integer less than or equal to x. Define a function f from non-negative reals into itself by

$$f(x) = x - [x]$$

Discuss the continuity of this function at every integer point.

Let $n \varepsilon N$. Then $f(n) = n - [n] = n - n = 0$.

$$\lim_{x \to n-} f(x) = \lim_{x \to n-} (x - [x]) = \lim_{x \to n-} x - \lim_{x \to n-} [x]$$

$$= n - (n - 1) = 1$$

$$\lim_{x \to n+} f(x) = \lim_{x \to n+} (x - [x])$$

$$= n - n = 0.$$

Hence the function is discontinuous at every integer point $n = 1, 2, 3, \ldots$. But observe that it is continuous at every other point.

(d) Let $f : R \to R$ be such that $f(x + y) = f(x) + f(y)$ and let f be continuous at $x = 0$. Prove that f is continuous for every $x \varepsilon R$.

Since $f(x + y) = f(x) + f(y)$, by taking $x = 0 = y$ we obtain $f(0) = 0$. Putting $y = -x$ we get $f(-x) = -f(x)$. Hence $f(x - y) = f(x) - f(y)$. Since f is continuous at $x = 0$, given $\varepsilon > 0$ there exists $\delta > 0$ such that

$$|f(x) - f(0)| < \varepsilon \text{ for } 0 < |x - 0| < \delta$$

i.e.,

$$|f(x)| < \varepsilon \text{ for } 0 < |x| < \delta$$

Therefore, if $x, y \varepsilon R$ are such that $|x - y| < \delta$, then

$$|f(x - y)| < \varepsilon.$$

i.e.,

$$|f(x) - f(y)| < \varepsilon.$$

This shows that f is continuous at every point.

(e) Show that the function $f: R \to R$ defined by

$$f(x) = \begin{cases} 1, & x \text{ irrational} \\ -1, & x \text{ rational} \end{cases}$$

is discontinuous everywhere.

Let a be any rational number. Then for every positive integer n, there exists an irrational number a_n such that $|a_n - a| < \frac{1}{n}$, and thus the sequence $\{a_n\}$ converges to a. Also

$$f(a_n) = 1 \text{ for all } n \text{ and so}$$

$$\lim_{n\to\infty} f(a_n) = 1.$$

But $f(a) = -1$ and thus

$$\lim_{n\to\infty} f(a_n) \neq f(a).$$

Therefore f is not continuous at any rational point. Next assume that b is any irrational number. Then for every positive integer n, there exists a rational number $\{b_n\}$ such that $|b_n - b| < \frac{1}{n}$ and hence $\lim_{n\to\infty} b_n = b$.

But $f(b_n) = -1$ for all n and so

$$\lim_{n\to\infty} f(b_n) = -1,$$

whereas $f(b) = 1$ and thus

$$\lim_{n\to\infty} f(b_n) \neq f(b).$$

Therefore f is not continuous at any irrational point.

(f) Show that the function $f: R \to R$ defined by

$$f(x) = \begin{cases} x, & x \text{ irrational} \\ -x, & x \text{ rational} \end{cases}$$

is continuous only at $x = 0$ and discontinuous at every other point.

To show the continuity of f at $x = 0$, let $\varepsilon > 0$ be given. Choose $0 < \delta < \varepsilon$. Let x be rational and $0 < |x| < \delta$. Then

$$|f(x) - f(0) = |-x| = |x| < \varepsilon$$

Let x be irrational and $0 < |x| < \delta$. Then

$$|f(x) - f(0)| = |x| < \varepsilon.$$

Hence in either case we obtain

$$|f(x) - f(0)| < \varepsilon \text{ whenever } 0 < |x| < \delta$$

and thus f is continuous at $x = 0$.

The discontinuity of f at every non-zero real number can be proved in a manner similar to that followed in the previous example and hence is left to the reader as an exercise.

(g) Let $f : R \to R$ be such that $f(x + y) = f(x) + f(y)$ and let f be continuous. Then show that $f(x) = x f(1)$ for all $x \in R$.

$$\text{Observe that } f(0) = 0 = 0 f(1).$$

Also it is true that $f(x) = -f(x)$.

Let x be a positive integer. Then

$$f(x) = \underbrace{f(1 + 1 + \ldots + 1)}_{x \text{ times}}$$

$$= \underbrace{f(1) + f(1) + \ldots + f(1)}_{x \text{ times}} \qquad \text{since } f(x + y) = f(x) + f(y)$$

$$= x f(1).$$

Let x be a negative integer. Then $x = -y$ where y is a positive integer. Now

$$f(x) = f(-y) = -f(y) = -y f(1) = x f(1).$$

Next let x be any rational number. Then x can be written in the form a/b where b is a positive integer, a any integer and a and b are coprime. Then

$$a f(1) = f(a) = f\left(\frac{a}{b} \cdot b\right)$$

$$= \underbrace{f\left(\frac{a}{b} + \frac{a}{b} + \ldots + \frac{a}{b}\right)}_{b \text{ times}}$$

$$= b f\left(\frac{a}{b}\right)$$

So
$$f(x) = f\left(\frac{a}{b}\right) = \frac{a}{b} f(1) = x f(1)$$

Finally, let x be any irrational number. Then there exists a sequence $\{x_n\}$ of rational numbers such that $\lim_{n\to\infty} x_n = x$. Since x_n is rational for each $n \in N$ we have

$$f(x_n) = x_n f(1) \text{ for all } n \in N.$$

Since f is continuous at x,

$$\begin{aligned} f(x) &= \lim_{n\to\infty} f(x_n) = \lim_{n\to\infty} x_n f(1) \\ &= \left(\lim_{n\to\infty} x_n\right) f(1) \\ &= xf(1). \end{aligned}$$

This proves the result for every real x.

(h) Let $f : R \to R$ be defined by

$$f(x) = \begin{cases} 1, & x \text{ rational} \\ 0, & x \text{ irrational} \end{cases}$$

Show that f is discontinuous at every point. This is left to the reader as an exercise.

(i) Let f: $R \to R$ be defined by

$$f(x) = \begin{cases} x, & x \text{ rational} \\ 0, & x \text{ irrational} \end{cases}$$

This is also left to the reader to show that f is continuous at $x = 0$ and discontinuous at every other point.

4.2.8. Definition

Let f be a real function defined on an open interval (a, b). The function f is said to be monotonically increasing on (a, b) if $a < x < y < b$ implies $f(x) \le f(y)$ and monotonically decreasing on (a, b) if $a < x < y < b$ implies $f(x) \ge f(y)$. The function f is said to be monotonic if either it is monotonically increasing or monotonically decreasing.

4.2.9. Theorem

Let f be a monotonic real function defined on (a, b). Then left-handed and right-handed limits of f exist at each point $x \in (a, b)$. To be precise, let f be monotonically increasing. Then $f(x+)$ and $f(x-)$ exist at each $x \in (a, b)$ and

$$\sup_{t \in (a,\, x)} f(t) = f(x-) \le f(x) \le f(x+) = \inf_{t \in (x,\, b)} f(t).$$

Further if $a < x < y < b$, then

$$f(x+) \le f(y-)$$

(Similar results hold if f is a monotonically decreasing function which is left to the reader to formulate and prove).

Proof: Let

$$A = \{f(t) \varepsilon R : t \varepsilon (a, x) \} \subset R\}.$$

Since f is monotonically increasing on (a, b), the set A is bounded above by the number $f(x)$. Therefore A has a supremum. Let $l = \sup A$. Clearly $l \le f(x)$. We now claim that $l = f(x-)$. To establish our claim, let $\varepsilon > 0$ be given. Since l is the supremum of the set A, there exists $\delta > 0$ such that $a < x - \delta < x$ and $l - \varepsilon < f(x - \delta) \le l$.

Further, since f is monotonically increasing,

$$f(x - \delta) \le f(t) \le l \text{ for } x - \delta < t < x.$$

Therefore we have

$$l - \varepsilon < f(x - \delta) \le f(t) \le l \text{ for } a < x - \delta < t < x$$

i.e.

$$|f(t) - l| < \varepsilon \text{ for} x - \delta < t < x.$$

In other words,

$$\lim_{t \to x-} f(t) = f(x-) = l = \sup_{t \varepsilon (a, x)} f(t).$$

It can be proved in a similar manner that

$$f(x) \le \lim_{t \to x+} f(t) = f(x+) = \inf_{t \varepsilon (x,\, b)} f(t).$$

Finally, assume that $a < x < y < b$. Then we have

$$f(x+) = \inf_{t \varepsilon (x,\, b)} f(t)$$

Since f is increasing and $x < y < b$,

$$\inf_{t \varepsilon (x,\, b)} f(t) = \inf_{t \varepsilon (x,\, y)} f(t).$$

Therefore

$$f(x+) = \inf_{t \varepsilon (x,\, y)} = f(t).$$

Similarly it can be shown that

$$f(y-) = \sup_{t \in (a, y)} f(t) = \sup_{t \in (x, y)} f(t).$$

Thus $f(x+) \leq f(y-)$ for $a < x < y < b$ and this completes the proof.

4.2.10. Theorem

Let f be monotonic on (a, b). Then the points of discontinuities of f at most form a countable set.

Proof: To be precise, let f be monotonically increasing. Let A denote the set of all points of (a, b) where f is not continuous. We have to show that A is countable. Since f is increasing it follows from the above theorem that for each $x \in (a, b)$,

$$f(x-) \leq f(x+).$$

Let $x \in A$. Then

$$f(x-) < f(x+).$$

Choose a rational number $r(x)$ such that

$$f(x-) < r(x) < f(x+).$$

Thus for each $x \in A$ we associate a rational number $r(x)$ which satisfies the above inequality. Let B denote the set of all such rationals. This correspondence between A and B is $1-1$. For, let $x_1 \neq x_2$. To be specific, say $x_1 < x_2$. This implies $f(x_1+) \leq f(x_2-)$. Therefore

$$f(x_1-) < r(x_1) < f(x_1+) \leq f(x_2-) < r(x_2) < f(x_2+),$$

which means that $r(x_1) < r(x_2)$, i.e., $r(x_1) \neq r(x_2)$. Hence there exists a 1–1 correspondence between A and B. But $B \subset Q$ and Q, the set of all rational numbers, is countable. Since any subset of a countable set is countable, it follows that B is countable. Hence A is countable and this completes the proof.

4.3 CONTINUITY AND COMPACTNESS

4.3.1. Theorem

The image of a compact subset of a metric space under a continuous mapping is compact.

Proof: Let (X, d_1) and (Y, d_2) be metric spaces and let $f: X \to Y$ be continuous. Let $K \subset X$ be compact. We have to show that $f(K)$ is compact. Let $\{G_i : i \varepsilon I\}$ be an open cover of $f(K)$. Then $\{f^{-1}(G_i) : i \varepsilon I\}$ is an open cover of K. Since K is compact, this open cover has a finite sub-cover. This means that there exist $G_1, G_2, \ldots, G_n$ such that

$$K = \bigcup_{k=1}^{n} f^{-1}(G_k) = f^{-1}\left(\bigcup_{k=1}^{n} G_k\right).$$

Therefore it follows that

$$f(K) \subset \bigcup_{k=1}^{n} G_k$$

Thus $f(K)$ has a finite sub-cover $\{G_1, G_2, \ldots, G_n\}$ and hence $f(K)$ is compact.

4.3.2. Corollary

A continuous real function defined on a compact metric space is bounded.

Proof: Let (X, d) be a compact metric space and let $f: X \to R$ be continuous. Then it follows from the above theorem that $f(X)$ is compact. But $f(X) \subset R$ and therefore by the Heine-Borel Theorem it is closed and bounded. Thus $f(X)$, the range of f, is a bounded set and hence the function f is bounded.

4.3.3. Maximum Value Theorem

Let (X, d) be a compact metric space and let $f: X \to R$ be continuous. Then f attains its bounds.

Proof: It follows from the previous corollary that f is bounded, Let

$$m = \inf_{x \varepsilon X} f(x) \text{ and } M = \sup_{x \varepsilon X} f(x).$$

We have to show that there are points c and d in X such that

$$f(c) = m \text{ and } f(d) = M.$$

Since X is compact, $f(X)$ is compact and hence it is closed and bounded. Since $f(X)$ is bounded, m and M exist. Since $f(X)$ is closed, both m and M are in $f(X)$. This means that there exist c and d in X such that

$$f(c) = m \text{ and } f(d) = M,$$

and this completes the proof.

4.3.4. Definition (Uniform Continuity)

Let (X, d_1) and (Y, d_2) be metric spaces and $f: X \to Y$. The function f is said to be uniformly continuous on X if for every $\varepsilon > 0$ there exists a $\delta > 0$. depending only on ε, such that

4.3.5.

$$d_2\,(f(x), f(y)) < \varepsilon \text{ for } d_1\,(x,y) < \delta.$$

4.3.6. Remark

On the surface it may appear to the reader that concept of continuity and that of uniform continuity are the same, but it should be noted that there exists an essential difference. Continuity is defined point-wise, whereas *uniform* continuity is a property defined on a set. If $x \in X$, then f is continuous at x if for a given $\varepsilon > 0$, there exists *a* $\delta > 0$ depending on ε and x such that 4.3.5 holds. If f is continuous at each point of X, then for a given $\varepsilon > 0$, each time we may get a different δ and it may not be possible to find only one δ which will satisfy 4.3.5 for each $x \in X$. But in case of uniform continuity of f, there exists only one $\delta > 0$, depending only on ε but not on any point such that 4.3.5. holds for points x and y in X.

Thus it follows from the above analysis that every uniformly continuous function is continuous, but the following example will show that the converse is not always true. We will soon observe that the converse holds if X, the domain of f, is compact.

4.3.7. Examples

(a) Shows that the function f defined by $f(x) = x^2$ is uniformly continuous on $[-1, 1]$.

Let $x, y \in [-1, 1]$. Then

$$\begin{aligned} |f(x) - f(y)| &= |x^2 - y^2| = |(x+y)\ (x-y)| \\ &= |x+y|\ |x-y| \end{aligned}$$

Given $\varepsilon > 0$ choose $\delta = \frac{\varepsilon}{2}$, so that, since $|x+y| \le 2$, whenever $|x-y| < \delta$ we have

$$|f(x) - f(y)| < 2 \cdot \frac{\varepsilon}{2} = \varepsilon.$$

Thus the required condition is satisfied with a choice of δ which is independent of points choosen and hence f is uniformly continuous on $[-1, 1]$.

(b) Show that the function defined by $f(x) = \frac{1}{x}$ is not uniformly continuous on $[0, 1]$.

The reader should observe that the function is continuous in $(0, 1]$. For uniform continuity given $\varepsilon > 0$ we must have $\delta > 0$ independent of choice of points x, y in $(0, 1]$ such that

$$\left|\frac{1}{x} - \frac{1}{y}\right| < \varepsilon \text{ when } |x - y| < \delta.$$

If $y = \delta$ this should also be true. But if $y = \delta$ we have $|x - \delta| < \delta$, i.e. $|x| < 2\delta$, whereas

$$\left|\frac{1}{x} - \frac{1}{\delta}\right| = \left|\frac{\delta - x}{\delta x}\right| \to \infty \text{ as } x \to 0.$$

4.3.8. Theorem

A continuous mapping defined on a compact metric space is uniformly continuous.

Proof: Let (X, d_1) and (Y, d_2) be metric spaces. Let $f : X \to Y$ be continuous and let X be compact. We have to show that f is uniformly continuous. Let $\varepsilon > 0$ be given. Since f is continuous on X, there exists $\delta_x > 0$ such that

$$d_2(f(x), f(y)) < \frac{\varepsilon}{2} \text{ for } d_1(x, y) < \delta_x.$$

Let G_x be the set defined by

$$G_x = \{ y \,\varepsilon\, X : d_1(x, y) < \frac{1}{2}\delta_x \}.$$

Let $\{G_{x_i} : i \,\varepsilon\, I\}$ be an open cover of X. Since X is compact, there exist finite number of points $x_1, x_2, \ldots, x_n$ in X such that

$$X \subset \bigcup_{k=1}^{n} G_{x_k}$$

Let

$$\delta = \frac{1}{2} \min \{\delta_{x_1}, \delta_{x_2}, \ldots, \delta_{x_n}\}.$$

Then $\delta > 0$ and is independent of x.

Let $x, y \varepsilon X$ such that $d_1(x, y) < \delta$. Then $x \varepsilon G_{x_m}$ for at least one $m = 1, 2, \ldots, n$. This shows that $d_1(x, x_m) < 1/2\ \delta\ x_m$. Hence $d_2(f(x), f(x_m)) < \frac{\varepsilon}{2}$. Also

$$d_1(x_m, y) \leq d_1(x_m, x) + d_1(x, y)$$
$$< \frac{1}{2}\delta x_m + \delta \leq \delta_{x_m}.$$

Therefore

$$d_2(f(x_m), f(y)) < \frac{\varepsilon}{2}.$$

Thus

$$d_2(f(x), f(y)) \leq d_2(f(x), f(x_m)) + d_2(f(x_m), f(y))$$
$$< \frac{\varepsilon}{2} + \frac{\varepsilon}{2} = \varepsilon.$$

So f is uniformly continuous and this completes the proof of the theorem.

4.4 CONTINUITY AND CONNECTEDNESS

4.4.1. Theorem

A continuous image of a connected metric space into another metric space is connected.

Proof: Let (X, d_1) and (Y, d_2) be metric spaces and let X be connected. Let $f : X \to Y$ be a continuous map. We have to show that $f(X)$ is connected. Assume, to the contrary, that $f(X)$ is disconnected. Then there exist disjoint nonempty open sets A and B in Y, both of which intersect $f(X)$ and

$$f(X) \subset A \cup B.$$

Since f is continuous and A and B are open, $f^{-1}(A)$ and $f^{-1}(B)$ are open sets in X. It is left to the reader to verify that $f^{-1}(A)$ and $f^{-1}(B)$ are nonempty, disjoint and

$$f^{-1}(A) \cup f^{-1}(B) = X.$$

Thus X is disconnected and this contradicts the hypothesis that it is connected. Thus the theorem is proved.

4.4.2. Intermediate Value Theorem

Let (X, d) be a connected metric space and let $f : X \to R$ be continuous. If $x_1, x_2 \varepsilon X$ and $f(x_1) < c < f(x_2)$, then there exists a point $x \varepsilon X$ such that $f(x) = c$.

Proof: Since X is connected, it follows from the previous theorem that $f(X)$ is connected. Since $f(X) \subset R$ it must be an interval. Since $f(x_1)$, $f(x_2) \varepsilon f(X)$ and $f(x_1) < c < f(x_2)$, it follows that $c \varepsilon f(X)$. Therefore there exists an $x \varepsilon X$ such that $f(x) = c$ and this completes the proof of the theorem.

4.5 CONVEX FUNCTIONS

In this section we shall discuss about convex functions and their relationship with continuity. The concept of convex functions plays a very important role in applications, particularly, in the theory of optimization, operations research, and economics. We shall first mention the definition of convex functions and discuss some examples.

Let I denote the closed interval $[a, b]$ on the real line R and let f: $I \to R$.. The function f is said to be convex on I if for $\lambda \in [0, 1]$ and $x, y \in I$,

$$f(\lambda x + (1 - \lambda) y) \leq \lambda f(x) + (1 - \lambda) f(y).$$

The function f is said to be strictly convex if strict inequality holds for $x \neq y$. f is said to be concave if for $\lambda \in [0, 1]$, $x, y \in I$

$$f(\lambda x + (1 - \lambda) y) \geq \lambda f(x) + (1 - \lambda) f(y)$$

and strictly concave if strict inequality holds for $x \neq y$.

The reader should be able to prove that the following functions are convex. In each case $f: I = [a, b] \to R$.

1. $f(x) = mx + c$, where $m, c \in R$ are fixed but arbitrary.
2. $f(x) = |x|$,
3. $f(x) = x^2 - 2x$
4. $f(x) = -x^{1/2}$, $[a, b] \subset R^+$
5. $f(x) = x^2$
6. $f(x) = e^x$, $[a, b] \subset R^+$

Theorem: Let $f : I \to R$ $g : I \to R$ be convex functions and let $\alpha \geq 0$. Then the functions $f + g$ and αf are convex.

The proof is left to the reader as an exercise. It is only to check that $f + g$ and αf satisfy the definition of convex function.

Recall that a function $f : I \to \mathbf{R}$ is said to be linear if for all x, $y \in I$ and $\alpha \in R$,

$$f(x+y) = f(x) + f(y),$$
$$f(\alpha x) = \alpha f(x).$$

$f : I \to R$ is said to be sublinear if for all x, $y \in$ I, $\alpha \geq 0$,

$$f(x+y) \leq f(x) + f(y),$$
$$f(\alpha x) = \alpha f(x).$$

Theorem: Every linear function is both convex and concave. A sublinear function is convex.

The proof is left to the reader as an exercise.

Theorem: Let $f : I \to I$ and $g : I \to R$ be convex and let g be nondecreasing. Then the composite function gf is convex.

Proof: Let x, $y \in I$ and $0 \leq \lambda \leq 1$. Then, since f is convex,

$$f(\lambda x) + (1 - \lambda) y)$$
$$\leq \lambda f(x) + (1 - \lambda) f(y).$$

Since g is nondecreasing and convex, we have

$$g [f(\lambda x + (1 - \lambda) y]$$
$$\leq g [\lambda f(x) + (1 - \lambda) f(y)]$$
$$\leq \lambda g (f(x) + (1 - \lambda) g (f(y)),$$

and this proves that gf is convex.

Theorem: If $f : I \to R$ and $g : I \to R$ are both nonnegative, nonincreasing [nondecreasing] and convex, then the function h defined by $h(x) = f(x) g(x)$ also has all these three properties.

Proof: We shall only prove the convexity of h which is nontrivial. Observe that for $x < y$,

$$[f(x) - f(y)] [g(y) - g(x)] \leq 0$$

This implies that

$$f(x) g(y) + f(y) g(x) \leq f(x) g(x) + f(y) g(y)$$

Now let $\alpha > 0$, $\beta > 0$ and $\alpha + \beta = 1$. Then

$$f(\alpha x + \beta y) g(\alpha x + \beta y)$$
$$\leq [\alpha f(x) + \beta f(y)] [\alpha g(x) + \beta g(y)]$$

$$= \alpha^2 f(x)\, g(x) + \alpha\ \beta\, [f(x)\, g(y) + f(y)\, g(x)] + \beta^2 f(y)\, g(y)$$

$$\leq \alpha^2 f(x)\, g(x) + \alpha\, \beta\, [f(x)\, g(x) + f(y)\, g(y)] + \beta^2 f(y)\, g(y)$$

$$= \alpha f(x)\, f(x) + \beta f(y)\, g(y)$$

and this completes the proof.

Theorem: A convex real function which is finite on a closed interval $[a, b]$ is bounded (from above and below).

Proof: Let f be convex and finite on $[a, b]$. Let $\max\{f(a), f(b)\} = M$. For any $z = \lambda a + (1 - \lambda) b$ in the interval $[a, b]$ we have

$$f(z) = f(\lambda a + (1 - \lambda)b) \leq \lambda f(a) + (1 - \lambda) f(b)$$

$$\leq \lambda M + (1 - \lambda) M = M$$

Thus f is bounded above by $M = \max\{f(a), f(b)\}$.

To see that f is bounded from below, observe that any arbitrary point can be, without any loss of generality, expressed as $\frac{a+b}{2} + t$.

Then we have

$$f\left(\frac{a+b}{2}\right) \leq \frac{1}{2} f\left(\frac{a+b}{2} + t\right) + \frac{1}{2} f\left(\frac{a+b}{2} - t\right)$$

Thus

$$f\left(\frac{a+b}{2} + t\right) \geq 2f\left(\frac{a+b}{2}\right) - f\left(\frac{a+b}{2} - t\right)$$

Since M is the upper bound we have

$$-f\left(\frac{a+b}{2} - t\right) \geq -M$$

So

$$f\left(\frac{a+b}{2} + t\right) \geq 2f\left(\frac{a+b}{2}\right) - M = m \text{ (say)}$$

This completes the proof.

Definition: Let $f : [a, b] \to R$. The function f is said to be Lipschitzian if there exists a constant $K > 0$ such that

$$|f(x) - f(y)| \leq K\, |x - y|$$

for any two points $x, y \in [a, b]$.

Observe that a Lipschitzian function defined on $[a, b]$ is uniformly continuous on $[a, b]$. For, let f be Lipschitzian and let $\in > 0$ be given.

Then there exists a $\delta = \epsilon /K$ such that $|x-y| < \delta$ implies $|f(x) - f(y)| < \epsilon$.

Theorem: Let $f: I \to R$ be convex. Then f is Lipschitzian on any closed interval $[a, b]$ contained in the interior I^0 of I. f is continuous on I^0.

Proof: Choose $\epsilon > 0$ in such a way that $[a-\epsilon, b+\epsilon] \subset I$. Let M and m be the upper and lower bounds, respectively, of f on $[a-\epsilon, b+\epsilon]$. If $x, y \in [a, b]$ and $x \neq y$, put

$$z = y + \frac{\epsilon}{|y-x|}(y-x), \quad \lambda = \frac{|y-x|}{\epsilon + |y-x|}$$

Then $z \in [a-\epsilon, b+\epsilon]$, $y = \lambda z + (1-\lambda)x$ and we have

$$\begin{aligned} f(y) &\leq \lambda f(z) + (1-\lambda) f(x) \\ &= \lambda [f(z) - f(x)] + f(x) \end{aligned}$$

$$\begin{aligned} f(y) - f(x) &\leq \lambda (M-m) < \frac{|y-x|}{\epsilon}(M-m) \\ &= K\,|y-x| \end{aligned}$$

where $K = \dfrac{M-m}{\epsilon}$. Since this is true for any $x, y \in [a, b]$ we conclude that f is Lipschitzian.

Since $[a, b] \subset I^0$ is an arbitrary closed interval in I^0, it follows that f is continuous in I^0 and this completes the proof.

Theorem: Let $f: I \to R$ be convex. Then for $a, b, x \in I$ and $x \in (a, b)$,

$$\frac{f(x)-f(a)}{x-a} \leq \frac{f(b)-f(a)}{b-a} \leq \frac{f(b)-f(x)}{b-x}$$

If f is strictly convex, then strict inequality holds.

Proof: Since f is convex we have

$$f(x) \leq \frac{b-x}{b-a} f(a) + \frac{x-a}{b-a} f(b).$$

Therefore it follows that

$$\begin{aligned} f(x) - f(a) &\leq \frac{a-x}{b-a} f(a) + \frac{x-a}{b-a} f(b) \\ &= \frac{x-a}{b-a}[f(b) - f(a)] \end{aligned}$$

This shows that the first inequality holds. The second one can be proved similarly. Also observe that if f is strictly convex, the above inequality is strict and hence the required inequality is strict.

EXERCISES

1. Answer 'Yes' or 'No'.

(i) Limit of $f(x)$ at $x = a$ is the same as limit of $f(x - h)$ at $h = 0$.

(ii) A continuous function is bounded.

(iii) A real function $f(x)$ of real x is continuous at a point if its left-hand limit is equal to its value at that point.

(iv) In R^2 it is not necessary that the limit of a real function exists for the continuity of the function at a point.

(v) A bounded function cannot be discontinuous at a point.

(vi) In R^2, a real function continuous along a curve may not be continuous along another curve.

(vii) In a metric space a real function defined on an isolated point is continuous at that point.

2. State which of the following are correct:

(i) $\lim\limits_{x \to \alpha + n} |[x]| = n$ for $0 < \alpha < 1$.

(ii) $f : X \to Y$ implies that f is continuous iff V is open in Y for U open in X where X and Y are metric spaces.

(iii) Only a continuous function can be monotonic.

(iv) A real function $f(x)$, x real is discontinuous at all points if

$$f(x) = \begin{cases} a & \text{for } x \in Z \\ 0 & \text{for } x \notin Z, \end{cases}$$

Z is the set of all integers.

(v) A continuous function maps a compact metric space into a metric space.

(vi) A discontinuous function cannot map, a compact metric space into a compact metric space.

3. Show that:

(i) $$\lim_{x \to 0+} \frac{1}{1 + e^{-1/2}} = 1$$

(ii) $$\lim_{x \to 0} \frac{x\, e^{1/x}}{1 + e^{1/x}} = 0$$

(iii) $\lim_{x \to 0} \dfrac{e^{1/x} - 1}{1 + e^{1/x}}$ does not exist.

(iv) $\lim_{x \to 0} \dfrac{e^{1/x} - 1}{e^{1/x} + 1}$ does not exist.

4. Investigate the continuity of each of the following functions at the indicated points:

(i) $$f(x) = \begin{cases} \dfrac{x\, e^{1/x}}{1 + e^{1/x}}, & x \neq 0 \\ 0 & x = 0 \end{cases} \quad \text{at } x = 0$$

(ii) $$f(x) = \begin{cases} \dfrac{e^{1/x^2}}{1 - e^{1/x^2}}, & x \neq 0 \\ 1 \quad , & x = 0 \end{cases} \quad \text{at } x = 0$$

(iii) $f(x) = x - |x|$ at $x = 0$

(iv) $$f(x) = \begin{cases} x^m \sin \dfrac{1}{x}, & x \neq 0 \text{ , } m > 0 \\ 0, & x = 0 \end{cases} \quad \text{at } x = 0$$

5. Define $f : R \to R$ by

$$f(x) = \begin{cases} 0, & x \quad \text{irrational} \\ \dfrac{1}{n}, & x = \dfrac{m}{n} \quad \text{(rational)} \end{cases}$$

Show that f is continuous at every irrational point and discontinuous at every rational point.

6. Show that

(i) $f : R \to R$ defined by

$$f(x) = x^2$$

is uniformly continuous in [0, 1] but not so on R.

(ii) $f : R \to R$ defined by

$$f(x) = x$$

is uniformly continuous in [0, 2].

7. Let (X, d) be a metric space. Define a map $f : X \to R$ by

$$f(x) = d(x, 0)$$

Show that f is continuous.

8. Let f be a uniformly continuous map of a metric space X into a metric space Y and let $\{x_n\}$ be a Cauchy sequence in X. Show that $\{f\{x_n\}\}$ is a Cauchy sequence in Y.

9. Let f be a continuous map defined on a metric X. Let

$$K_f = \{x \,\varepsilon\, X : f(x) = 0\}$$

Show that K_f is a closed set.

10. Let f be a uniformly continuous function defined on a bounded set $A \subset R$. Show that f is bounded on A.

11. Study the continuity of $f : R \to R$ defined by

$$f(x) = \begin{cases} 0 & x \text{ irrational} \\ \dfrac{1}{q}, & x \text{ rational, } x = \dfrac{p}{q}, \ (p, q) = 1 \end{cases}$$

on $(0, 1)$.

12. If $a_n \to l$ as $n \to \infty$, show that

$$(a_1 \ a_2 \ \ldots \ a_n)^{\frac{1}{n}} \to l \text{ as } n \to \infty.$$

[Use the continuity of log function.]

5 Differentiability

5.1 DEFINITION AND PROPERTIES

5.1.1. Definition

Let $[a, b]$ be a closed interval on the real line R and let f be a real-valued function defined on $[a, b]$. Let $x_0 \varepsilon [a, b]$. For $x \varepsilon (a, b)$ and $x \neq x_0$ let

$$\lim_{x \to x_0} \frac{f(x) - f(x_0)}{x - x_0}$$

if it exists, or equivalently for $h \neq 0$

$$\lim_{h \to 0} \frac{f(x_0 + h) - f(x_0)}{h}$$

if it exists, be denoted by $f'(x_0)$.

Thus, given a function $f : [a, b] \to R$ we obtain a function f' which is defined on $A \subset [a, b]$ at all points of which either one (and hence both) of the above limits exist. The function f' thus obtained is called the derivative of the function f. If f' is defined at a point x_0, then f is said to be differentiable at that point and $f'(x_0)$ is called the differential coefficient of f at x_0. If f is differentiable at every point on a set $A \subset [a, b]$, then f is said to be differentiable on A.

Given $x_0 \varepsilon [a, b]$ if

$$\lim_{x \to x_0^+} \frac{f(x) - f(x_0)}{x - x_0}$$

exists, then f is said to have a right–handed derivative at x_0 or is said to be right differentiable at x_0 and this is denoted by $f'_+(x_0)$. Similarly, if

$$\lim_{x \to x_0^-} \frac{f(x) - f(x_0)}{x - x_0}$$

exists, then f has a left–handed derivative at x_0 and the above limit is written as $f'_-(x_0)$.

5.1.2. Remark

(a) f is differentiable at a point $x_0 \in (a, b)$ if both $f'_+(x_0)$ and $f'_-(x_0)$ exist and are equal, and in this case we write

$$f'_+(x_0) = f'_-(x_0) = f'(x_0)$$

(b) Let f: $[a, b] \to R$. If f has a derivative at a. then it is a right-handed one. Similarly if $f'(b)$ exists, then it is a left-handed derivative.

(c) Let f: $[a, b] \to [a, b]$ be defined by

$$f(x) = x.$$

Then $f'(x) = 1$ for all $x \in [a, b]$

(d) Let f be a constant, i.e., let f: $[a, b] \to R$ be such that for some $c \in R$,

$$f(x) = c \text{ for all } x \in [a, b].$$

Then $f'(x) = 0$ for all $x \in [a, b]$.

5.1.3. Examples

(a) f: $[0, 1] \to R$ defined by

$$f(x) = x^2$$

is differentiable on $[0, 1]$.

Let $x_0 \in (0, 1)$. Then

$$f'(x_0) = \lim_{x \to x_0} \frac{x^2 - x_0^2}{x - x_0} = \lim_{x \to x_0} (x + x_0) = x_0 + x_0 = 2x_0$$

$$f'(0) = \lim_{x \to 0+} \frac{x^2 - 0}{x - 0} = \lim_{x \to 0+} \frac{x^2}{x} = \lim_{x \to 0} x = 0$$

$$f'(1) = \lim_{x \to 1-} \frac{x^2 - 1}{x - 1} = \lim_{x \to 1-} (x + 1) = 2$$

Thus f is differentiable on $[0, 1]$; it is also continuous there.

(b) The function f defined by

$$f(x) = |x|$$

is not differentiable at the origin

$$f'_-(0) = \lim_{x \to 0-} \frac{|x| - |0|}{x - 0} = \lim_{x \to 0-} \frac{|x|}{x}$$

$$= \lim_{x \to 0-} \frac{-x}{x} = -1.$$

$$f'_+(0) = \lim_{x \to 0+} \frac{|x| - |0|}{x - 0} = \lim_{x \to 0+} \frac{x}{x}$$

$$= \lim_{x \to 0+} \frac{x}{x} = 1.$$

Thus f is not differentiable at 0. But this function is continuous at $x = 0$.

(c) The function f defined by

$$f(x) = \begin{cases} x \sin 1/x & (x \neq 0) \\ 0 & (x = 0) \end{cases}$$

is not differentiable at the origin.

In this case

$$\frac{f(x) - f(0)}{x - 0} = \frac{x \sin \frac{1}{x}}{x} = \sin \frac{1}{x}$$

and this does not tend to a limit as $x \to 0$. Hence this is not differentiable at 0, but is continuous there.

(d) The function f defined by

$$f(x) = \begin{cases} x^2 \sin \frac{1}{x} & (x \neq 0) \\ 0 & (x = 0) \end{cases}$$

is differentiable (and also continuous) at 0. Here

$$f'(0) = \lim_{x \to 0} \frac{f(x) - f(0)}{x - 0} = \lim_{x \to 0} \frac{x^2 \sin \frac{1}{x}}{x}$$

$$= \lim_{x \to 0} x \sin \frac{1}{x} = 0.$$

5.1.4. Theorem

If a function $f: [a, b] \to R$ is differentiable at a point $x_0 \ \varepsilon \ [a, b]$, then it is continuous there. The converse of this statement is not true.

Proof: Let f be differentiable at a point $x_0 \ \varepsilon \ [a, b]$.

Then we have

$$f'(x_0) = \lim_{x \to x_0} \frac{f(x) - f(x_0)}{x - x_0}.$$

Now

$$\lim_{x \to x_0} [f(x) - f(x_0)] = \lim_{x \to x_0} \left[\frac{f(x) - f(x_0)}{x - x_0} (x - x_0) \right]$$

$$= \left[\lim_{x \to x_0} \frac{f(x) - f(x_0)}{x - x_0} \right] \left[\lim_{x \to x_0} x - x_0 \right]$$

$$= f'(x_0) \cdot 0 = 0.$$

Thus

$$\lim_{x \to x_0} f(x) = f(x_0)$$

and hence f is continuous at x_0.

That the converse is not true follows from the examples 5.1.3 (b) and (c).

Several partinent questions now arise, e.g., one would naturally ask whether the sum, scalar product, product, quotient and compositions of differentiable functions are differentiable. These questions are answered in the following theorems.

5.1.5. Theorem

Let $f, g : [a, b] \to R$ be differentiable at a point $x_0 \ \varepsilon \ [a, b]$ and let $\alpha \ \varepsilon \ R$. Then $f + g$, αf, fg and $\frac{f}{g}$ (if $g(x_0) \neq 0$) are differentiable at x_0 and

(i) $(f + g)'(x_0) = f'(x_0) + g'(x_0)$,

(ii) $(\alpha f)'(x_0) = \alpha f'(x_0)$,

(iii) $(fg)'(x_0) = f'(x_0)\, g(x_0) + f(x_0)\, g'(x_0)$,

(iv) $$\left(\frac{f}{g}\right)'(x_0) = \frac{g(x_0)\, f'(x_0) - f(x_0)\, g'(x_0)}{[g(x_0)]^2}$$

Proof: We shall prove only (iii) and (iv). For the proof of (iii) let $h = fg$. Then

$$h(x) - h(x_0) = f(x)\,[g(x) - g(x_0)] + g(x_0)\,[f(x) - f(x_0)].$$

So that

$$\frac{h(x) - h(x_0)}{x - x_0} = f(x)\,\frac{g(x) - g(x_0)}{x - x_0} + g(x_0)\,\frac{f(x) - f(x_0)}{x - x_0}$$

Letting $x \to x_0$ we have $f(x) \to f(x_0)$ and

$$h'(x_0) = f(x_0)\, g'(x_0) + g(x_0)\, f'(x_0).$$

$$= f'(x_0)\, g(x_0) + f(x_0)\, g'(x_0).$$

To prove (iv) let $h = \frac{f}{g}$. Then

$$\frac{h(x) - h(x_0)}{x - x_0} = \frac{1}{g(x)\, g(x_0)}$$

$$\left[g(x)\,\frac{f(x) - f(x_0)}{x - x_0} - f(x_0)\,\frac{g(x) - g(x_0)}{x - x_0} \right]$$

Letting $x \to x_0$ we have $g(x) \to g(x_0)$ and hence

$$h'(x_0) = \frac{1}{[g(x_0)]^2}\,[g(x_0)\, f'(x_0) - f(x_0)\, g'(x_0)].$$

5.1.6. Remarks

(a) If f and g are differentiable at x_0, then $f - g$ is also differentiable at x_0 and

$$(f - g)'(x_0) = f'(x_0) - g'(x_0)$$

(b) If f is differentiable at x_0 and $f(x_0) \neq 0$, then $\frac{1}{f}$ is differentiable at x_0 and

$$\left(\frac{1}{f}\right)'(x_0) = \frac{-f'(x_0)}{[f(x_0)]^2}.$$

(c) Let f be defined by

$$f(x) = x^n.$$

If $n \geq 1$, then $f'(x) = n x^{n-1}$ for any x.

If $n < 1$, then $f'(x) = n x^{n-1}$ for $x \neq 0$

(d) Any polynomial

$$f(x) = a_0 + a_1 x + a_2 x^2 + .. + a_n x^n$$

where $a_0, a_1, a_2, \ldots, a_n$ are all real numbers, is differentiable at any point.

(e) Any rational function defined by

$$\frac{f(x)}{g(x)}$$

is differentiable except at the points where $g(x) = 0$.

5.1.7. Theorem

Suppose a real function f is defined and continuous on $[a, b]$, differentiable at some point $x_0 \in [a, b]$, another real function g is defined and continuous on an interval I containing the range of f and g is differentiable at $f(x_0)$. Then the composition of f and g, gof, is differentiable at x_0 and

$$(gof)'(x_0) = g'(f(x_0)) f'(x_0).$$

Proof: Let $y_0 = f(x_0)$. Since g is differentiable at y_0,

$$g(y) - g(y_0) = (y - y_0)\ [g'(y_0) + \vartheta(y)]$$

where $y \in I$ and $\vartheta(y) \to 0$ as $y \to y_0$.

Since f is differentiable at x_0,

$$f(x) - f(x_0) = (x - x_0)\ [f'(x_0) + u(x)]$$

where $x \in [a, b]$ and $u(x) \to 0$ as $x \to x_0$.

Let $y = f(x)$ and let $h = gof$. Then

$$\begin{aligned} h(x) - h(x_0) &= g(f(x)) - g(f(x_0)) \\ &= [f(x) - f(x_0)]\ [g'(y_0) + \vartheta(y)] \\ &= (x - x_0)\ [f'(x_0) + u(x)]\ [g'(y_0) + \vartheta(y)] \end{aligned}$$

Now if $x \neq x_0$,

$$\frac{h(x) - h(x_0)}{x - x_0} = [\,g'(y_0) + \vartheta(y)\,]\,[f'(x_0) + u(x)]$$

Letting $x \to x_0$ we see that, since f is continuous at x_0,

$$y = f(x) \to f(x_0) = y_0 \text{ and that}$$

$$h'(x_0) = g' f(x_0) \cdot f'(x_0).$$

This completes the proof of the theorem.

5.2 MEAN VALUE THEOREMS

5.2.1. Rolle's Theorem

If $f: [a, b] \to R$ is continuous on $[a, b]$, differentiable on (a, b) and $f(a) = f(b)$, then there exists a point $c \in (a, b)$ such that $f'(c) = 0$.

Proof: Since f is continuous on the compact set $[a, b]$ it follows from the Maximum Value Theorem that f is bounded and attains its bounds. In other words, if

$$m = \inf_{x \in [a,b]} f(x) \text{ and } M = \sup_{x \in [a,b]} f(x),$$

then there exist points c and d in $[a, b]$ such that

$$f(c) = m \text{ and } f(d) = M.$$

If $m = M$, then f is constant and hence $f'(x) = 0$ for all $x \in [a, b]$. Thus the result holds trivially. Assume that $m \neq M$. Since $f(a) = f(b)$, both m and M cannot be equal to $f(a)$, at least one of them, say $m = f(c)$, is different from $f(a)$. Thus $c \neq a$, similarly it can be shown that $c \neq b$, and hence $c \in (a, b)$. We now claim that $f'(c) = 0$. We have

$$f(c + h) - f(c) \leq 0 \text{ for any } h.$$

$$\frac{f(c+h) - f(c)}{h} \begin{cases} \leq 0 \text{ if } h \geq 0 \\ \geq 0 \text{ if } h \leq 0 \end{cases}$$

Since $f'(c)$ exists, this is impossible unless $f'(c) = 0$ and thus our claim is established.

5.2.2. Lagrange Mean Value Theorem

If $f: [a, b] \to R$ is continuous on $[a, b]$ and differentiable on (a, b), then there exists a point $c \in (a, b)$ such that

$$\frac{f(b) - f(a)}{b - a} = f'(c).$$

Proof: Define a function F by

$$F(x) = f(x) - \frac{f(b) - f(a)}{b - a}\, x$$

F is defined and continuous on $[a, b]$, differentiable on (a, b) and $F(a) = F(b)$. Thus F satisfies the conditions of Rolle's Theorem. Therefore, there exists a point $c \in (a, b)$ such that $F'(c) = 0$. But for $x \in (a, b)$,

$$F'(x) = f'(x) - \frac{f(b) - f(a)}{b - a}$$

Hence we have

$$f'(c) = \frac{f(b) - f(a)}{b - a}$$

and this completes the proof.

5.2.3. Cauchy Mean Value Theorem

If $f, g : [a, b] \to R$ are

(i) continuous on $[a, b]$,

(ii) differentiable on (a, b),

(iii) $g'(x) \neq 0$ for any $x \in (a, b)$,

then there exists a point $c \in (a, b)$ such that

$$\frac{f(b) - f(a)}{g(b) - g(a)} = \frac{f'(c)}{g'(c)}$$

Proof: Define a function F by

$$F(x) = f(x) - \frac{f(b) - f(a)}{g(b) - g(a)}\, g(x)$$

F satisfies the conditions of Rolle's Theorem and hence there exists a point $c \in (a, b)$ such that $F'(c) = 0$. Now for $x \in (a, b)$,

$$F'(x) = f'(x) - \frac{f(b) - f(a)}{g(b) - g(a)}\, g'(x)$$

Therefore

$$\frac{f'(c)}{g'(c)} = \frac{f(b) - f(a)}{g(b) - g(a)}$$

and this completes the proof.

5.2.4. Theorem

Let $f : (a, b) \to R$ be differentiable on (a, b). Then for all $x \,\varepsilon\, (a, b)$,

(i) $f'(x) \geq 0 \Rightarrow f$ is monotonically increasing,

(ii) $f'(x) \leq 0 \Rightarrow f$ is monotonically decreasing,

(iii) $f'(x) = 0 \Rightarrow f$ is a constant.

Proof: Let $x_1, x_2 \,\varepsilon\, (a, b)$. Observe that f satisfies the conditions of Lagrange Mean Value Theorem in the interval $[x_1, x_2]$. Therefore there exists a point $x \,\varepsilon\, (x_1, x_2)$ such that

$$\frac{f(x_2) - f(x_1)}{x_2 - x_1} = f'(x).$$

All the conclusions of the theorem now follow from the above equation.

5.2.5. Theorem

Let $f : [a, b] \to R$ be differentiable. Then f' assumes all the values between $f'(a)$ and $f'(b)$, more precisely, if $f'(a) < c < f'(b)$ then there exists a point $x_0 \,\varepsilon\, (a, b)$ such that $f'(x_0) = c$.

Proof: Define a function F by

$$F(x) = f(x) - cx \text{ for all } x \,\varepsilon\, [a, b].$$

Then $F'(x) = f'(x) - c$.

Observe that $F'(a) = f'(a) - c < 0$, so there exists $x_1 \,\varepsilon\, (a, b)$ such that $F(x_1) < F(a)$. Further $F'(b) = f'(b) - c > 0$. Hence there exists $x_2 \,\varepsilon\, (a, b)$ such that $F(x_2) < F(b)$. Since f is differentiable on $[a, b]$ F is also differentiable and hence is continuous on the compact set $[a, b]$. Hence F attains its bounds. To be precise, if $m = \inf F(x)$ then there exists a point x_0 such that $F(x_0) = m$. Now $x_0 \neq a$, for otherwise $m = F(x_0) = F(a) > F(x_1)$ and this contradicts the definition of m. Similarly $x_0 \neq b$, showing that $x_0 \,\varepsilon\, (a, b)$. We now claim that $F'(x_0) = 0$. For if $F'(x_0) < 0$, there would exist a point x for which $F(x) < F(x_0) = m$, which would contradict the definition of m. A similar contradiction will be obtained if we assume that $F'(x_0) > 0$. Thus $F'(x_0) = f'(x_0) - c = 0$, showing that

$$f'(x_0) = c$$

and this completes the proof.

5.2.6. Corollary (Darboux Theorem)

Let $f: [a, b] \rightarrow R$ be differentiable on $[a, b]$ and let $f'(a), f'(b)$ have opposite signs. Then there exists a point $x_0 \in (a, b)$ such that $f'(x_0) = 0$.

Proof: $f'(a)$ and $f'(b)$ have opposite signs. More precisely, let $f'(a) < 0$ and $f'(b) > 0$. Then 0 belongs to the interval $(f'(a), f'(b))$. Therefore by above theorem there exists $x_0 \in (a, b)$ such that $f'(x_0) = 0$.

The reader is advised to prove this result independently by arguing as in the proof of previous theorem.

5.2.7. Theorem

For $a, b \in R$ show that

$$\frac{|a+b|}{1+|a+b|} \leq \frac{|a|}{1+|a|} + \frac{|b|}{1+|b|}$$

Proof: Define a function $f: R^+ \rightarrow R^+$ by

$$f(x) = \frac{x}{1+x}$$

where R^+ denotes the set of all nonnegative real numbers. We have

$$f'(x) = \frac{1}{(1+x)^2} > 0 \text{ for all } x \in R^+.$$

Therefore it follows that f is monotonically increasing. Further, for $a, b \in R$, $|a+b| \leq |a| + |b|$. Hence

$$f(|a+b|) \leq f(|a| + |b|)$$

or,
$$\frac{|a+b|}{1+|a+b|} \leq \frac{|a|+|b|}{1+|a|+|b|}$$

$$\leq \frac{|a|}{1+|a|+|b|} + \frac{|b|}{1+|a|+|b|}$$

$$\leq \frac{|a|}{1+|a|} + \frac{|b|}{1+|b|}.$$

5.2.8. Theorem (L' Hospitals Rule for Indeterminate $\frac{0}{0}$ form)

Let f and g be two real functions such that

(i) both are differentiable and $g'(x) \neq 0$ for all x in an open interval $(a - \delta, a + \delta)$, except possibly at a,

(ii) $f(a) = 0 = g(a)$,

(iii) $\lim_{x \to a} \frac{f'(x)}{g'(x)}$ exists.

Then $\lim_{x \to a} \frac{f(x)}{g(x)}$ exists and

$$\lim_{x \to a} \frac{f(x)}{g(x)} = \lim_{x \to a} \frac{f'(x)}{g'(x)}.$$

Proof: Let x belong to the open interval $(a - \delta, a + \delta)$. Then f and g satisfy the conditions of Cauchy's Mean Value Theorem in the interval $[a, x]$. Therefore there exists a point $c \in (a, x)$ such that

$$\frac{f(x) - f(a)}{g(x) - g(a)} = \frac{f'(c)}{g'(c)}$$

Since $f(a) = 0 = g(a)$, we have,

$$\frac{f(x)}{g(x)} = \frac{f'(c)}{g'(c)}.$$

Since $c \in (a, x)$ as $x \to a$ we have $c \to a$. Therefore

$$\lim_{x \to a} \frac{f(x)}{g(x)} = \lim_{x \to a} \frac{f'(c)}{g'(c)} = \lim_{c \to a} \frac{f'(c)}{g'(c)} = \lim_{x \to a} \frac{f'(x)}{g'(x)}$$

and this completes the proof.

5.2.9. Example

$$\lim_{x \to 0} \frac{\sin x}{x} = \lim_{x \to 0} \frac{\cos x}{1}$$

$$= \frac{\lim_{x \to 0} \cos x}{1} = \frac{1}{1} = 1.$$

The readers have spent a considerable amount of time in the intermediate classes doing exercises in computation of limits of functions by using L'Hospital Rule which are in $\frac{0}{0}$, $0 \times \infty$, $\infty - \infty$, 0^0, 1^∞ and ∞^0 forms. Therefore we will not present any solutions of such examples in this text. A number of such problems will, however, be given in the exercises at the end of this chapter for readers to work out.

5.3 TAYLOR'S THEOREM

Let f be a real function differentiable on an interval $[a, b]$ and let f' be differentiable also. Then we write f'' to denote the derivative of f'. Similarly if f'' is differentiable we write f''' to denote its derivative. In this way we write f^n to denote the derivative of f^{n-1} if it exists, for any positive integer n. The derivative f^n, if it exists, is called the nth order derivative of f, in particular, f', f'' and f''' are respectively called the first, second and third order derivatives.

5.3.1. Taylor's Theorem

Let f be a real function such that

(i) f^{n-1} is continuous in $[a, a+h]$,

(ii) f^n exists in $(a, a+h)$.

Then for a given positive integer m, there exists a number $\theta \, \varepsilon \, (0, 1)$ such that

5.3.2.

$$f(a+h) = f(a) + hf'(a) + \frac{h^2}{2!} f''(a) + ..$$

$$+ \frac{h^{n-1}}{(n-1)!} f^{n-1}(a) + \frac{h^n (1-\theta)^{n-m}}{(n-1)!\, m} f^n(a + \theta h).$$

Proof: Let

$$S_n = f(a) + hf'(a) + \frac{h^2}{2!} f''(a) + \ldots + \frac{h^{n-1}}{(n-1)!} f^{n-1}(a)$$

Define a function F by

$$F(x) = f(x) + (a+h-x) f'(x) + \frac{(a+h-x)^2}{2!} f''(x) + \ldots$$

$$+ \frac{(a+h-x)^{n-1}}{(n-1)!} f^{n-1}(x) + A(a+h-x)^m,$$

where

5.3.3.

$$Ah^m = f(a+h) - S_n.$$

It is left to the reader to verify that F is continuous in $[a, a + h]$, differentiable in $(a, a+h)$ and $F(a) = (a+h)$. Therefore Rolle's Theorem is applicable for F in the interval $[a, a + h]$ and hence there exists a number $\theta \in (0, 1)$ such that

$$F'(a + \theta h) = 0$$

But

$$F'(x) = \frac{(a+h-x)^{n-1}}{(n-1)!} f^n(x) - mA(a+h-x)^{m-1}.$$

Therefore

$$0 = F'(a + \theta h).$$

$$= \frac{h^{n-1}(1-\theta)^{n-1}}{(n-1)!} f^n(a+\theta h) - mA(1 - \theta)^{m-1} h^{m-1}$$

$$\text{or, } A = \frac{h^{n-m}(1-\theta)^{n-m}}{(n-1)!\, m} f^n(a+\theta h).$$

Putting this value of A in 5.3.3 we get 5.3.2 and this completes the proof of the theorem.

5.3.4. Remark

Let $R_n = \dfrac{h^n(1-\theta)^{n-1}}{(n-m)!} f^m(a+\theta h)$

R_n is called the remainder of the Taylor's expansion (5.3.2), two special cases of which are well known and commonly used.

If $m = 1$, then

$$R_n = \frac{h^n(1-\theta)^{n-1}}{(n-1)!} f^n(a+\theta h)$$

and this is called the Cauchy form of the remainder of Taylor's expansion (5.3.2).

If $m = n$, then

$$R_n = \frac{h^n}{n!} f^n(a+\theta h)$$

and this is known as Lagrange's form of remainder.

5.3.5. Maclaurin's Theorem

Let f be such that

(i) f^{n-1} is continuous in $[0, h]$,

(ii) f^{n} exists in $(0, h)$.

Then, for any given positive integer m, there exists a number $\theta \varepsilon (0, 1)$ such that

$$f(h) = f(0) + hf'(0) + \frac{h^2}{2!}f''(0) + \ldots$$

$$+ \frac{h^{n-1}}{(n-1)!}f^{n-1}(0) + \frac{h^n(1-\theta)^{n-m}}{(n-1)!\,m}f^n(\theta h)$$

This is a special case of Taylor's Theorem, the case for which $a = 0$, and thus the proof is exactly similar to that of Taylor's Theorem. The reader should prove this result independently.

The Cauchy's form of remainder of Maclaurin's expansion is given by

$$R_n = \frac{h^n(1-\theta)^{n-1}}{(n-1)!}f^n(\theta h)$$

and the Langrange's form is given by

$$R_n = \frac{h^n}{n!}f^n(\theta h).$$

5.3.6. Taylor's Infinite Series

If a real function f is such that,

(i) it possesses derivatives of all orders in $[a, a+h]$, and

(ii) the Taylor's remainder $R_n \to 0$ as $n \to \infty$.

$$f(a+h) = f(a) + hf'(a) + \frac{h^2}{2!}f''(a) + \ldots$$

i.e., the infinite series on the right of the above equation converges to $f(a+h)$.

Proof: If the derivatives of all orders of f exist in $[a, a+h]$, then f satisfies the hypothesis of Taylor's theorem.

Hence

$$f(a+h) = S_n + R_n$$

where S_n and R_n have been defined in 5.3.1 and 5.3.4 respectively. If $\lim_{n\to\infty} R_n = 0$, then

$$\lim_{n\to\infty}(S_n + R_n) = \lim_{n\to\infty} S_n + \lim_{n\to\infty} R_n$$

$$= \lim_{n\to\infty} S_n$$

$$= f(a) + hf'(a) + \frac{h^2}{2!}f''(a) + \dots \frac{h^n}{n!} f^n(a) + ..$$

Therefore we have

$$f(a+h) = f(a) + hf'(a) + \frac{h^2}{2!} f''(a) + ...$$

5.3.7. Maclaurin's Infinite Series

If f has derivatives of all orders in $[0, h]$ and if Maclaurin's remainder $\to 0$ as $n \to \infty$, then

$$f(h) = f(0) + hf'(0) + \frac{h^2}{2!} f''(0) + ...$$

By using the above theorems the readers must have derived Infinite Series expansions of various functions such as : e^x, $\log(1 + x)$, $\cos x$, $\sin x$, $\tan x$, $\cos^{-1} x$, $\sin^{-1} x$ etc. in their calculus courses. We shall therefore solve only some problems here and provide number of them in the exercises for readers to workout for themselves.

5.3.8. Examples

(a) Expand $f(x) = e^x$ in a Maclaurin's series,

Let $f(x) = e^x$. Then for any $n \varepsilon N$, $f^n(x) = e^x$. Also f and all its derivatives are continuous for every $x \varepsilon R$. Thus we can expand f in Maclaurin's series. Consider the Lagrange remainder.

$$R_n = \frac{x^n}{n!} f^n(\theta x) = \frac{x^n}{n!} e^{\theta x}, \ 0 < \theta < 1 .$$

$$\lim_{n\to\infty} R_n = \lim_{n\to\infty} \frac{x^n}{n!} e^{\theta x} = \left(\lim_{n\to\infty} \frac{x^n}{n!}\right) e^{\theta x} = 0$$

since

$$\lim_{n\to\infty} \frac{x^n}{n!} = 0 \text{ and } e^{\theta x} \text{ is bounded.}$$

Therefore, for all $x \varepsilon R$,

$$e^x = 1 + x + \frac{x^2}{2!} + \frac{x^3}{3!} + \ldots$$

(b) Find Maclaurin's series for $f(x) = \cos x$.

$$\text{Let } f(x) = \cos x. \text{ Then } f'(x) = \cos\left(\frac{n\pi}{2} + x\right)$$

In this case f and all its derivatives are continuous for every $x \varepsilon R$. Now

$$R_n = \frac{x^n}{n!} f^n(\theta x) = \frac{x^n}{n!} \cos\left(\frac{n\pi}{2} + \theta x\right)$$

$$|R_n| = \left|\frac{x^n}{n!} \cos\left(\frac{n\pi}{2} + \theta x\right)\right| = \left|\frac{x^n}{n!}\right| \left|\cos\left(\frac{n\pi}{2} + \theta x\right)\right|$$

$$\leq \left|\frac{x^n}{n!}\right|$$

Therefore

$$\lim_{n \to \infty} |R_n| = 0 \text{ for all } x \varepsilon R \text{ and thus}$$

$$\cos x = 1 - \frac{x^2}{2!} + \frac{x^4}{4!} - \frac{x^6}{6!} + \ldots$$

(c) Maclaurin's series for $f(x) = \sin x$ is given by

$$\sin x = x - \frac{x^3}{3!} + \frac{x^5}{5!} - \ldots.$$

(d) Find the Maclaurin's series for $\log(1 + x)$.

Let $f(x) = \log(1 + x)$. $f(x)$ possesses derivatives of all orders for $1 + x > 0$ i.e., for $x > -1$ and for $x > -1$,

$$f^n(x) = \frac{(-1)^{n-1}(n-1)!}{(l+x)^n}.$$

If R_n denotes the Lagrange form of remainder, we have

$$R_n = \frac{x^n}{n!} f^n(\theta x) = (-1)^{n-1} \frac{1}{n}\left(\frac{x}{1 + \theta x}\right)^n.$$

For $0 \le x \le 1$, $\left(\dfrac{x}{1 + \theta x}\right)^n$ is positive and less than 1 for $n = 1, 2, \ldots$.
Also since

$$\frac{1}{n} \to 0 \text{ as } n \to \infty, \; R_n \to 0.$$

For $-1 < x < 0$, let R_n denote the Cauchy's form of remainder. Thus

$$R_n = \frac{x^n}{(n-1)!}(1-\theta)^{n-1} f^n(\theta x)$$

$$= (-1)^{n-1} x^n \cdot \frac{1}{1+\theta x}\left(\frac{1-\theta}{1+\theta x}\right)^{n-1}.$$

Note that $\dfrac{1-\theta}{1+\theta x}$ is positive and less than 1 and

$$\frac{1}{1+\theta x} < \frac{1}{1-|x|}.$$

Also $x^n \to 0$ as $n \to \infty$ and hence $R_n \to 0$. Further,

$$f^n(0) = (-1)^{n-1}(n-1)!$$

Therefore

$$\log(1+x) = x - \frac{x^2}{2} + \frac{x^3}{3} - \frac{x^4}{4} + \ldots$$

for $-1 < x \le 1$.

(e) Find the Maclaurin's series for $(1 + x)^m$.

For any real m, $(1 + x)^m$ possesses continuous derivatives of all orders for $1 + x > 0$ i.e., for $x > -1$ and

$$f^n(x) = m(m-1)(m-2) \ldots (m-n+1)(1+x)^{m-n}$$

If m is a positive integer, then $f^n(x) = 0$ for $n > m$. Hence $R_n = 0$ for $n > m$ and thus $(1 + x)^m$ has only $(m + 1)$ terms.

We now discuss the case when m is not a positive integer. Let R_n denote the Cauchy's form of remainder. Then

$$R_n = \frac{x^n}{(n-1)!}(1-\theta)^{n-1} f^n(\theta x)$$

$$= x^n \frac{m(m-1)\ldots(m-n+l)}{(n-1)!}\left(\frac{1-\theta}{1+\theta x}\right)^{n-1}(1+\theta x)^{n-1}$$

Let $-1 < x < 1$. We have

$$0 < \left(\frac{1-\theta}{1+\theta x}\right) < 1.$$

Also

$$(1+\theta x)^{m-1} \begin{cases} < 2^{m-1} \text{ if } (m-1) \text{ is positive} \\ \leq (1-|x|)^{m-1} \text{ if } (m-1) \text{ is negative.} \end{cases}$$

Further

$$\frac{m(m-1)\ldots(m-n+1)}{(n-1)!} x^n \to 0 \text{ for } |x| < 1.$$

Hence $R_n \to 0$ for $|x| < 1$.

$$f^n(0) = m(m-1)\ldots(m-n+1).$$

Hence for $|x| < 1$,

$$(1+x)^m = 1 + mx + \frac{m(m-1)}{2!}x^2 + \ldots..$$

Theorem: Let $f : I \to R$ be convex. Then f has both a right and a left derivative, at every point of I^0, the interior of I, and f'_-, f'_+ are non-decreasing on I^0. If $c \in I^0$, then we have

$$f'_-(c) \leq f'_+(c),$$

$$f(x) \geq f(c) + f'_-(c)\,(x-c),$$

$$f(x) \geq f(c) + f'_+(c)\,(x-c)$$

for all $x \in I$.

Proof: Let $c \in I^0$ and let $[a, b] \subset I$ be such that $c \in (a, b)$. Then we have

$$\frac{f(c)-f(a)}{c-a} \leq \frac{f(x)-f(c)}{x-c}$$

for $x \in (c, b)$. It also follows that the function

$$x \to \frac{f(x)-f(c)}{x-c}$$

is nondecreasing on (c, b). Hence the right derivative

$$f'_+(c) = \lim_{x \to c_+} \frac{f(x)-f(c)}{x-c}$$

exists. In a similar manner it can be shown that the left derivative also exists.

If $a < c < d < b$ then for a sufficiently small positive h we have

$$\frac{f(c)-f(c-h)}{h} \leq \frac{f(c+h)-f(c)}{h} \leq \frac{f(d)-f(d-h)}{h}$$

Letting $h \to 0$ we obtain

$$f'_-(c) \leq f'_+(c) \leq f'_-(d).$$

Remark: Using the above theorem we now give an alternative and simpler proof of the fact that if $f: I \to R$ is convex, then it is Lipschitzian on any closed interval $[a, b] \subset I^0$.

For this choose the points c and d in I such that $c < a < b < d$ and this choice is possible. Then by the above theorem we have

$$f'_+(a) \leq f'_+(x) \leq \frac{f(x)-f(y)}{x-y} \leq f'_-(y) \leq f'_-(b)$$

whenever $a \leq x < y \leq b$. It now follows that

$$|f(x)-f(y)| \leq K\,|x-y|$$

where

$$K = \max\left\{|f'_+(a)|\ ,\ |f^{\pm}(b)|\right\}$$

and this completes the proof.

Theorem: Let $f: I \to R$ be convex. Then

(i) f'_- is left continuous and f'_+ is right continuous on I^0, and

(ii) the set of all points where f is not differentiable is at most countable.

Proof: (i) Since $f: I \to R$ is convex, it is continuous on I^0. Therefore, for all $x, y, z, \in I^0$ we have

$$\frac{f(y)-f(x)}{y-x} = \lim_{z \to x^+} \frac{f(y)-f(z)}{y-z} \geq \lim_{z \to x^+} f'_+(z)$$

for $x < z < y$. Letting $y \to x_+$ we obtain

$$f'_+(x) \geq \lim_{z \to x+} f'_+(z).$$

Since f'_+ is nondecreasing, we have

$$f'_+(x) \leq \lim_{z \to x+} f'_+(z)$$

Thus it follows that

$$f'_+(x) = \lim_{z \to x+} f'_+(z)$$

This shows that f'_+ is right continuous . In a similar manner, it can be shown that f'_- is left continuous.

(ii) Let $x, y, z \in I$ and let $x < y < z$. Then it follows that

$$f'_+(x) \le f'_-(y) \le f'_+(z)$$

Suppose that f'_+ is continuous at y. Then we have

$$f'_+(y) = \lim_{x \to y-} f'_+(x) \lim_{z \to y^+} = f'_+(z) = f'_-(y)$$

This shows that f is differentiable at y. Thus the points of I^0 where f is not differentiable are those points where the nondecreasing function f'_+ has a jump and there are only countably many such jumps. This completes the proof of (ii).

Theorem: Let I be open and let $f: I \to R$ be twice differentiable. Then f is convex if and only $f''(x) \ge 0$ for all $x \in I$.

Proof: Let f be convex. Then, since I is open, f' is nondecreasing on I. Hence $f''(x) \ge 0$ for all $x \in I$.

Conversely, let $f''(x) \ge 0$ for all $x \in I$. Let $x, y, \in I$, $x < y$ and $0 < \lambda < 1$. Then by the Mean-Value Theorem for differential calculus there exist ζ_1, ζ_2, $x < \zeta_1 < \lambda x + (1-\lambda)y < \zeta_2 < y$ and there exists ζ_3 with $\zeta_1 < \zeta_3 < \zeta_2$ such that

$$f(\lambda x + (1-\lambda)y) - \lambda f(x) - (1-\lambda) f(y)$$
$$= \lambda\, [f(\lambda x + (1-\lambda)y) - f(x)] + (1-\lambda)\, [f(\lambda x + (1-\lambda)y) - f(y)]$$
$$= \lambda(1-\lambda)(y-x) f'(\zeta_1) + (1-\lambda)\lambda(x-y) f'(\zeta_2)$$
$$= \lambda(1-\lambda)\ (y-x)(\zeta_1-\zeta_2) f''(\zeta_3) \le 0$$

Thus f is convex.

It should be noted that if f is strictly convex, then $f''(x) > 0$ for all $x \in I$, but the converse is not always true. For example, the function f defined by

$$f(x) = x^4$$

is strictly convex on R but $f''(0) = 0$.

EXERCISES

1. Answer 'Yes' or 'No'.

(i) The function $f(x)$, where

$$f(x) = \begin{cases} 0, & x \varepsilon N \\ 1, & x \not\varepsilon N \end{cases}$$

is not differentiable at integer points.

(ii) The function

$$f(x) = a_0 + a_1 |x| + a_2 |x|^2 + a_3 |x|^3 + .. + a_n |x|^n$$

is differentiable at all points for all real $a_0, a_1\ a_2, a_3, \ldots, a_n$.

(iii) The function

$$f(x) = a_0 + a_1 |x|^2 + a_2 |x|^4 + \ldots + a_n |x|^{2n}$$

is twice differentiable at $x = 0$.

2. Suppose $f(x)$ is defined in a neighbourhood of x and $f(x)$ is differentiable twice. Show that

$$\lim_{h \to 0} \frac{f(x+h) + f(x-h) - 2f(x)}{h^2} = f''(x)$$

3. Let $f: R \to R$ be defined by

$$f(x) = \begin{cases} 0, & x \le 0 \\ x, & x > 0 \end{cases}$$

Show that f is continuous but not differentiable at $x = 0$

4. Let $f: R^+ \to R$ be defined by

$$f(x) = \begin{cases} x, & 0 \le x < 1 \\ 1, & x \ge 1 \end{cases}$$

Show that $f(1)$ does not exist. Is f continuous at $x = 1$.

5. Discuss the differentiability of f defined by

$$f(x) = \begin{cases} x, & 0 \le x \le 1 \\ 1, & x > 1 \end{cases}$$

at $x = 1$.

6. Show that the function f defined by

$$f(x) = \begin{cases} x^3, & \sin \frac{1}{x}, \ x \ne 0 \\ 0, & x = 0 \end{cases}$$

has a derivative at $x = 0$ and further $f'(x)$ is continuous at $x = 0$.

7. Show that the following functions are differentiable at $x = 0$

(i) $f(x) = x|x|$ (ii) $f(x) = x^2|x|$

8. Examine Rolle's Theorem and Lagrange's Mean Value Theorem for

$$f(x) = |x| \text{ for } x \,\varepsilon\, [-1, 1].$$

9. Verify Rolle's theorem in $[a, b]$ for

$$f(x) = (x - a)^m (x - b)^n$$

where m and n are positive integers.

10. Verify Lagrange's Mean Value Theorem for

$$f(x) = x(x-1)(x-2)$$

in $\left[0, \frac{1}{2}\right]$

11. Examine Lagrange's Mean Value Theorem for

$$f(x) = \frac{1}{x} \text{ in } [-1, 1]$$

12. If $f(x) = |x|$ and $g(x) = 2|x|$, then $f'(0)$ and $g'(0)$ do not exist but $\lim_{x \to 0} \frac{f(x)}{g(x)}$ exists and

$$\lim_{x \to 0} \frac{f(x)}{g(x)} = \lim_{x \to 0} \frac{f'(x)}{g'(x)}$$

13. Find Maclaurin's infinite series expansion for

(i) $f(x) = \log(1 + x)$ for $[-1, 1]$

(ii) $f(x) = (1 + x)^m$ for any real constant m

14. Assuming the validity of expansion, show that

(i) $\log \sec x = \frac{1}{2} x^2 + \frac{1}{12} x^4 + \ldots$

(ii) $e^x \cos x = 1 + x - \frac{2x^3}{3!} - \frac{2^2 x^4}{4!} - \ldots$

15. Evaluate by using L'Hospital's rule:

(i) $\lim_{x \to 0} \frac{x - \tan x}{x^3}$

(ii) $\lim_{x \to 0} \frac{x e^x - \log(1 + x)}{x^2}$

(iii) $\lim\limits_{x \to 0} \dfrac{(1+x)^{1/x} - e}{x}$

(iv) $\lim\limits_{x \to 0} \left(\dfrac{1}{x^2} - \dfrac{1}{\sin^2 x} \right)$

(v) $\lim\limits_{x \to 0} \dfrac{x - \sin x}{x^3}$

(vi) $\lim\limits_{x \to 0} \dfrac{x - \log(1+x)}{1 - \cos x}$

(vii) $\lim\limits_{x \to \pi/2} (\sin x)^{\tan x}$

(viii) $\lim\limits_{x \to \infty} x \tan \dfrac{1}{x}$

(ix) $\lim\limits_{x \to 0} \dfrac{\log(1+x^3)}{\sin^3 x}$

(x) $\lim\limits_{x \to \infty} \dfrac{x^3}{e^x}$.

6 Riemann - Stieltjes Integral

6.1 RIEMANN INTEGRAL

6.1.1. Definition

Let $[a, b]$ be any closed and bounded interval in R. A subdivision or a partition Δ of $[a, b]$ is any finite set $\Delta \subset [a, b]$ of the form

$$\Delta = \{ a = x_0, x_1, x_2, \ldots, x_{n-1}, x_n = b\}$$

where

$$a = x_0 < x_1 < x_2 < \ldots < x_{n-1} < x_n = b.$$

We denote the family of all subdivisions of $[a,b]$ by $D[a,b]$.

Write

$$\Delta x_i = x_i - x_{i-1}, \ (i = 1, 2, \ldots, n).$$

Let f be a bounded real function defined on $[a, b]$. Let

$$M = \sup_{x \varepsilon [a,b]} f(x) \text{ and } m = \inf_{x \varepsilon [a,b]} f(x).$$

Since f is bounded on $[a, b]$, f is bounded in each of the subintervals $[x_{i-1}, x_i]$ corresponding to each subdivision $\Delta \varepsilon D[a, b]$, let

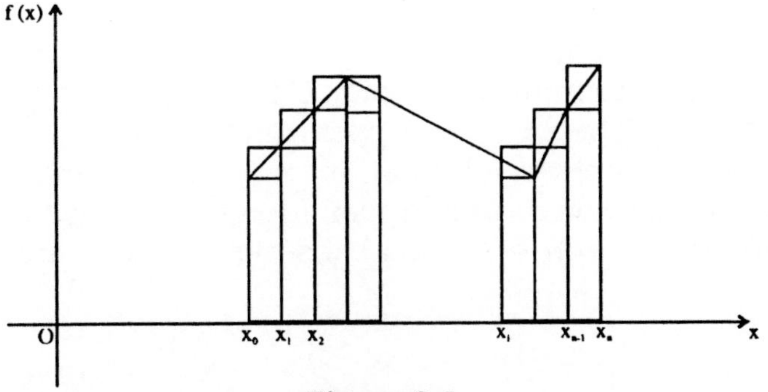

Figure 6.1

$$M = \sup_{x \varepsilon [x_{i-1}, x_i]} f(x) \text{ and } m_i = \inf_{x \varepsilon [x_{i-1}, x_i]} f(x).$$

Let

$$U(\Delta, f) = \sum_{i=1}^{n} M_i \Delta x_i, \; L(\Delta, f) = \sum_{i=1}^{n} m_i \Delta x_i$$

$U(\Delta, f)$ and $L(\Delta, f)$ are respectively called the upper and lower Darboux sums of f in $[a, b]$ corresponding to the subdivision Δ.

Observe that if $f(x) \geq 0$ for all $x \varepsilon [a, b]$, then the lower sum is numerically equal to the sum of the areas of the inscribed rectangles and the upper sum is equal numerically to the sum of the circumscribed rectangles.

For the definition observe that for any $\Delta \varepsilon D[a, b]$,

$$m \leq m_i \leq M_i < M$$

and so $$m \sum_{i=1}^{n} \Delta x_i \leq \sum_{i=1}^{n} m_i \Delta x_i \leq \sum_{i=1}^{n} M_i \Delta x_i \leq M \sum_{i=1}^{n} \Delta x_i$$

Hence

$$m(b-a) \leq L(\Delta, f) \leq U(\Delta, f) \leq M(b-a).$$

Put

$$\int_a^{-b} f dx = \inf_{\Delta \varepsilon D[a, b]} U(\Delta, f)$$

$$\int_{-a}^{b} f dx = \sup_{\Delta \varepsilon D[a, b]} L(\Delta, f)$$

$\int_a^{-b} f dx$ and $\int_{-a}^{b} f dx$ are respectively called the upper and lower integrals of f on $[a, b]$. Because of the above inequality $L(\Delta, f)$ is bounded above and hence has a supremum, similarly $U(\Delta, f)$ is bounded below and so has an infimum. Therefore the upper and lower integrals are defined for any bounded function f defined on any closed bounded interval $[a, b]$. Although the lower and upper integrals always exist for any bounded function, but we shall soon observe that they may not always be equal.

6.1.2. Definition

A bounded function f is said to be Riemann integrable on $[a, b]$ if the lower and upper integrals of f on $[a, b]$ are equal. The common value is called the integral of f and written as

$$\int_a^b f(x)dx \text{ or } \int_a^b fdx.$$

It should be noted that the Riemann integral of f, if it exists, is numerically equal to the area of the region bounded by the portion of the curve represented by the function with the ordinates at $x = a$, $x = b$ and the interval $[a, b]$.

6.1.3. Definition

Let Δ and Δ^* be two partitions of $[a, b]$. We say that Δ^* is a refinement of Δ if $\Delta \subset \Delta^*$.

Given two partitions Δ_1 and Δ_2 of $[a, b]$,

$$\Delta^* = \Delta_1 \cup \Delta_2.$$

is a common refinement of Δ_1 and Δ_2.

6.1.4. Theorem

(i) Let Δ^* be a refinement of Δ. Then

$$L(\Delta, f) \le L(\Delta^*, f) \le U(\Delta^*, f) \le U(\Delta, f)$$

(ii) If Δ_1 and Δ_2 are any two partitions, then

$$L(\Delta_1, f) \le U(\Delta_2, f)$$

(iii) $\int_{-a}^{b} f\,dx \le \int_{a}^{-b} f\,dx$

Proof: (i) Suppose that Δ^* contains only one extra point above that of Δ. Let this point be x^* and $x_{i-1} < x^* < x_i$. Since f is bounded on $[a, b]$, it is bounded on each subinternal. Let m_i, w_1, w_2 denote the infima of $f(x)$ on $[x_{i-1}, x_i]$, $[x_{i-1}, x^*]$ and $[x^*, x_i]$ respectively. Then $w_1 \ge m_i$ and $w_2 \ge m_i$.

Thus

$$L(\Delta, f) - L(\Delta^*, f) = m_i (x_i - x_{i-1}) - w_1 (x^* - x_{i-1}) - w_2 (x_i - x^*) \leq 0$$

or

$$L(\Delta, f) \leq L(\Delta^*, f)$$

On the same lines, we can show that

$$U(\Delta^*, f) \leq U(\Delta, f)$$

for any partition Δ we have that $L(\Delta, f) \leq U(\Delta, f)$

Hence

$$L(\Delta, f) \leq L(\Delta^*, f) \leq U(\Delta^*, f) \leq U(\Delta, f)$$

(ii) Let Δ_1 and Δ_2 be any two partitions and Let $\Delta = \Delta_1 \cup \Delta_2$ be their common refinement. Then from above

$$L(\Delta_1, f) \leq L(\Delta, f) \leq U(\Delta, f)$$

and

$$L(\Delta, f) \leq U(\Delta, f) \leq U(\Delta_2, f)$$

Hence

$$L(\Delta_1, f) \leq U(\Delta_2, f)$$

(iii) As Δ_1 and Δ_2 are arbitrary in (ii), we have

$$\sup_{\Delta \, \varepsilon \, D[a,b]} L(\Delta, f) \leq \inf_{\Delta \, \varepsilon \, D[a,b]} U(\Delta, f)$$

Hence

$$\int_{-a}^{b} f dx \leq \int_{a}^{-b} f dx$$

6.1.5. Theorem

A bounded function f is Riemann integrable on $[a, b]$ if and only if for every $\varepsilon > 0$ there exists a partition Δ such that

$$U(\Delta, f) - L(\Delta, f) < \epsilon$$

Proof: If f is Riemann integrable on $[a, b]$ then

$$\int_{-a}^{b} f dx = \int_{a}^{-b} f dx$$

Also as

6.1.6.

$$\int_{-a}^{b} f dx = \sup_{\Delta \varepsilon D} L(\Delta, f) \text{ and } \int_{a}^{-b} f dx = \inf_{\Delta \varepsilon D} U(\Delta, f)$$

there exists a partition Δ_1, such that

$$L(\Delta_1, f) + \frac{\varepsilon}{2} > \int_{-a}^{b} f dx$$

Similarly, for some partition Δ_2, we can have

$$U(\Delta_2, f) - \frac{\varepsilon}{2} < \int_{a}^{-b} f dx$$

Now, let Δ be a refinement of Δ_1 and Δ_2. Then

$$\begin{aligned} U(\Delta, f) &\leq U(\Delta_2, f) \\ &< \int_{a}^{-b} f dx + \frac{\epsilon}{2} \\ &= \int_{-a}^{b} f dx + \frac{\epsilon}{2} \\ &< L(\Delta_1, f) + \frac{\epsilon}{2} + \frac{\epsilon}{2} \\ &\leq L(\Delta, f) + \varepsilon \end{aligned}$$

Hence

$$U(\Delta, f) - L(\Delta, f) < \varepsilon$$

Conversely, let for every $\varepsilon > 0$ there is a partition Δ such that

6.1.7.

$$U(\Delta, f) - L(\Delta, f) < \epsilon$$

But from (6.1.6),

$$L(\Delta, f) \leq \int_{-a}^{b} f dx$$

and

$$\int_{a}^{-b} f dx \leq U(\Delta, f)$$

Therefore

$$L(\Delta, f) \leq \int_{-a}^{b} f dx \leq \int_{a}^{-b} f dx \leq U(\Delta, f)$$

The inequality (6.1.7) implies that

$$\int_{a}^{-b} f dx - \int_{-a}^{b} f dx < \varepsilon$$

As ε is arbitrary,

$$\int_{a}^{-b} f dx = \int_{-a}^{b} f dx$$

That is, f is Riemann integrable on $[a, b]$.

6.1.8. Theorem

Let f be Riemann integrable on $[a, b]$ and let c be such that $a < c < b$. Then

$$\int_{a}^{b} f dx = \int_{a}^{c} f dx + \int_{c}^{b} f dx$$

Proof: We have

$$[a, b] = [a, c] \cup [c, b]$$

Let Δ_1 and Δ_2 be some partitions of $[a, c]$ and $[c, b]$ respectively and $\Delta = \Delta_1 \cup \Delta_2$.

Then

$$L(\Delta, f) = L(\Delta_1, f) + L(\Delta_2, f)$$

or
$$\sup_{\Delta \varepsilon D[a,\, b]} L(\Delta f) = \sup_{\Delta_1 \varepsilon D[a,\, c]} L(\Delta_1, f) + \sup_{\Delta_2 \varepsilon D[c,\, b]} L(\Delta_2, f)$$

Hence

$$\int_{-a}^{b} f dx = \int_{-a}^{c} f dx + \int_{-c}^{b} f dx$$

Similarly, we can prove

$$\int_{a}^{-b} f dx = \int_{a}^{-c} f dx + \int_{c}^{-b} f dx$$

The integrability of f proves the theorem.

6.1.9. Theorem

If a function f is continuous on $[a, b]$, then it is Riemann integrable on $[a, b]$.

Proof: The interval $[a, b]$ on R is a compact set. Therefore continuity of f on $[a, b]$ implies uniform continuity of the function on $[a, b]$. Therefore, for a given $\varepsilon > 0$ we can find *a* $\delta > 0$ such that

$$|f(x) - f(y)| < \frac{\epsilon}{b-a} \text{ for } |x - y| < \delta$$

where $x, y \, \varepsilon \, [a, b]$

Further we take points $x_i \, \varepsilon \, [x_{i-1}, x_i]$ and $y_i \, \varepsilon \, [x_{i-1}, x_i]$ such that

$$x_i = \sup_{x \, \varepsilon \, (x_{i-1},\, x_i)} f(x) \text{ and } y_i = \inf_{x_i \, \varepsilon \, (x_{i-1},\, x_i)} f(x)$$

Thus we can write

$$|f(x_i) - f(y_i)| < \frac{\epsilon}{b-a} \text{ for } |x_i - y_i| < \delta$$

or
$$M_i - m_i < \frac{\epsilon}{b-a} \text{ for } |x_i - y_i| < \delta$$

Multiplying above inequality by Δx_i and taking summation we have

$$\sum_{i=1}^{n} (M_i - m_i) \, \Delta x_i < \frac{\epsilon}{b-a} \sum_{i=1}^{n} \Delta x_i$$

or

6.1.10.

$$U(\Delta, f) - L(\Delta, f) < \varepsilon$$

Inequality 6.1.10 proves that f is Riemann integrable.

6.1.11. Theorem

Every bounded monotonic function f is Riemann integrable. on $[a, b]$.

Proof: We consider the case of a monotonic increasing function. We construct a partition Δ such that $\Delta x_i < \dfrac{\epsilon}{f(b) - f(a)}$ for all $i = 1, 2, 3, \ldots n$ and $\varepsilon > 0$.

We have

$$U(\Delta, f) - L(\Delta, f) = \sum_{i=1}^{n} (M_i - m_i)\ \Delta x_i$$

But

$$M_i = f(x_i) \text{ and } m_i = f(x_{i-1})$$

Therefore

$$U(\Delta, f) - L(\Delta, f) = \sum_{i=1}^{n} [f(x_i) - (f(x_{i-1})]\ \Delta x_i$$

$$< \frac{\epsilon}{f(b) - f(a)} \sum_{i=1}^{n} [f(x_i) - f(x_{i-1})]$$

$$= \epsilon$$

Hence f is Riemann integrable on $[a, b]$.

6.2 RIEMANN STIELTJES INTEGRALS

6.2.1. Definition

Let $[a, b]$ be a bounded closed interval in R. Let α be monotonically increasing and let f be bounded, each defined on $[a, b]$. For each $\Delta \varepsilon D[a, b]$ write

$$\Delta \alpha_i = \alpha(x_i) - \alpha(x_{i-1}), \quad i = 1, 2, \ldots, n$$

and let

$$M = \sup_{x \varepsilon [a, b]} f(x), \quad m = \inf_{x \varepsilon [a, b]} f(x)$$

$$M_i = \sup_{x \in [x_{i-1}, x_i]} f(x), \quad m_i = \inf_{x \in [x_{i-1}, x_i]} f(x).$$

Define

$$L(\Delta, f, \alpha) = \sum_{i=1}^{n} m_i \Delta \alpha_i, \quad U(\Delta, f, \alpha) = \sum_{i=1}^{n} M_i \Delta \alpha_i$$

These are respectively called the *lower* and *upper sums*.

As in the case of Riemann integral one can see that for each $\Delta \varepsilon D[a, b]$.

$$m[\alpha(b) - \alpha(a)] \leq L(\Delta, f, \alpha) \leq U(\Delta, f, \alpha) \leq M[\alpha(b) - \alpha(a)]$$

We define

$$\overline{\int_a^b} f d\alpha = \inf_{\Delta \varepsilon D[a,b]} U(\Delta, f, \alpha)$$

$$\underline{\int_a^b} f d\alpha = \sup_{\Delta \varepsilon D[a,b]} L(\Delta, f, \alpha)$$

These are respectively called *upper* and *lower Riemann-Stieltjes* integrals and these exist because of the boundedness of $L(\Delta, f, \alpha)$ and $U(\Delta, f, \alpha)$ which follows from the above inequality.

If the upper and lower integrals of f are equal, then we say that f is Riemann-Stieltjes integrable and we denote the common value by

$$\int_a^b f d\alpha \text{ or } \int_a^b f(x)\, d\alpha(x).$$

We write $R(\alpha)$ to denote the set of all Riemann-Stieltjes integrable functions f with respect to α. A Riemann integral is a special case of Riemann-Stieltjes integral, the case for which $\alpha(x) = x$.

6.2.2. Theorem

(i) Let Δ^* be a refinement of Δ. Then

$$L(\Delta, f, \alpha) \leq L(\Delta^*, f, \alpha) \leq U(\Delta^*, f, \alpha) \leq U(\Delta, f, \alpha)$$

(ii) If Δ_1 and Δ_2 are any two partitions, then

$$L(\Delta_1, f, \alpha) \leq U(\Delta_2, f, \alpha)$$

(iii) $\int_{-a}^{b} f d\alpha \leq \int_{a}^{-b} f d\alpha.$

The statement (i) just means that by adding more points to a partition the lower sum increases but the upper sum decreases. Similarly (ii) says that every lower sum does not exceed any upper sum.

Proof: (i) Suppose that Δ^* contains only one extra point beyond that of Δ. Let this point be denoted by x^* and let $x^* \varepsilon (x_{i-1}, x_i)$. Since f is bounded on $[a, b]$, it is bounded on each of the subintervals $[x_{i-1}, x_i]$, $[x_{i-1}, x^*]$ and $[x^*, x_i]$. Let m_i, w_1, w_2 respectively denote the infimum of $f(x)$ on $[x_{i-1}, x_i]$. $[x_{i-1}, x^*]$ and $[x^*, x_i]$. Clearly $w_1 \geq m_i$, $w_2 \geq m_i$. Now

$$L(\Delta, f, \alpha) - L(\Delta^*, f, \alpha)$$

$$= m_i [\alpha(x_i) - \alpha(x_{i-1})] - w_1 [\alpha(x^*) - \alpha(x_{i-1})]$$

$$- w_2 [\alpha(x_i) - \alpha(x^*)]$$

$$= (m_i - w_1) [\alpha(x^*) - \alpha(x_{i-1})] + (m_i - w_2) [\alpha(x_i) - \alpha(x^*)] \leq 0.$$ If Δ^* contains k points more than those of Δ, then we repeat this process k times and see that

$$L(\Delta, f, \alpha) \leq L(\Delta^*, f, \alpha).$$

It follows from the definition and as has been observed earlier

$$L(\Delta^*, f, \alpha) \leq U(\Delta^*, f, \alpha).$$

The proof of the last inequality in (1) can be done in a manner similar to that of the first one.

(ii) Let Δ_1 and Δ_2 be any two partitions and let $\Delta = \Delta_1 \cup \Delta_2$ be their common refinement. Then from (i) it follows that

$$L(\Delta_1, f, \alpha) \leq L(\Delta, f, \alpha) \leq U(\Delta, f, \alpha) \leq U(\Delta_2, f, \alpha).$$

Therefore

$$L(\Delta_1, f, \alpha) \leq U(\Delta_2, f, \alpha).$$

(iii) Let Δ_1 and Δ_2 be any two partitions. Then

$$L(\Delta_1, f, \alpha) \leq U(\Delta_2, f, \alpha).$$

Hence $$\int_{-a}^{b} f d\alpha = \sup_{\Delta \, \varepsilon \, D[a, b]} L(\Delta, f, \alpha) \leq U(\Delta, f, \alpha)$$

Now $$\int_{-a}^{b} f d\alpha \leq \inf_{\Delta \,\varepsilon\, D[a,\, b]} U(\Delta, f, \alpha) = \int_{a}^{-b} f d\alpha$$

and this completes the proof.

6.2.3. Examples

(a) The function $f : R \to R$ defined by

$$f(x) = \begin{cases} 0, & x \text{ rational} \\ 1, & x \text{ irrational} \end{cases}$$

is not Riemann integrable on any interval.

Consider the interval $[a, b] \subset R$ and a partition

$$\Delta = \{ x_0 = a, x_1, x_2, \ldots, x_{n-1}, x_n = b \}.$$

Then

$$\begin{aligned} U(f, \Delta) &= \sum_{i=1}^{n} \Delta x_i \\ &= 1\,\Delta x_1 + 1\Delta x_2 + \ldots + 1\Delta x_n \\ &= b - a. \end{aligned}$$

Therefore

$$\begin{aligned} \int_{a}^{-b} f dx &= \inf U(f, \Delta) = \inf (b - a). \\ &= b - a. \end{aligned}$$

$$\begin{aligned} L(f, \Delta) &= \sum_{i=1}^{n} m_i \Delta x_i \\ &= 0\,\Delta x_1 + 0\,\Delta x_2 + \ldots + 0\,\Delta x_n = 0 \end{aligned}$$

Hence $$\int_{-a}^{b} f dx = \sup\ L(f, \Delta) = 0$$

Thus $$\int_{a}^{-b} f dx = b - a \neq 0 = \int_{-a}^{b} f dx$$

and so f is not Riemann integrable.

(b) Let α be an increasing function defined on an interval $[a, b]$ and continuous at $x_0 \in [a, b]$. Let a function f be defined on $[a, b]$ by

$$f(x) = \begin{cases} 0, & x \neq x_0 \\ 1, & x = x_0. \end{cases}$$

Prove that $f \in R(\alpha)$ on $[a, b]$ and $\int_a^b f d\alpha = 0$.

Let Δ be a partition of $[a, b]$, that is,

$$\Delta = \{ x_0 = a, x_1, x_2, \ldots, x_{n-1}, x_n = b\}.$$

Since $x_0 \in [a, b]$, it must belong to some subinterval, say Δx_i where $i \in \{1, 2, \ldots, n\}$. It is given that α is continuous at x_0. Therefore, for a given $\varepsilon > 0$ there exists a $\delta > 0$ such that

$$\Delta \alpha_i = \alpha(x_i) - \alpha(x_{i-1}) < \varepsilon, \text{ whenever } \Delta x_i < \delta.$$

Let Δ be such a partition that $\max\limits_{j \in \{1, 2, \ldots, n\}} \Delta x_j < \delta$.

Then $\Delta x_i < \delta$ and so $\Delta \alpha_i < \varepsilon$. Now

$$U(\Delta, f, \alpha) = \Delta \alpha_1.$$

and $L(\Delta, f, \alpha) = 0$.

Since ε is arbitrary,

$$\overline{\int_a^b} f d\alpha = \inf U(\Delta, f, \alpha) = 0.$$

Also
$$\underline{\int_a^b} f d\alpha = \sup L(\Delta, f, \alpha) = 0.$$

Thus $f \in R(\alpha)$ and $\int_a^b f d\alpha = 0$.

(c) If α is bounded and monotonically increasing on $[a, b]$ and if k is a constant, then show that $k \in R(\alpha)$ and

$$\int_a^b k \, d\alpha = k[\alpha(b) - \alpha(a)]$$

If $\alpha(x) = x$ for all $x \varepsilon [a, b]$, then

$$\int_a^b k dx = k(b-a).$$

For any partition Δ of $[a, b]$,

$$L(\Delta, k, \alpha) = k\Delta\alpha_1 + k\Delta\alpha_2 + \ldots + k\Delta\alpha_n$$

$$= k(\Delta\alpha_1 + \Delta\alpha_2 + \ldots \Delta\alpha_n)$$

$$= k\ \{[\alpha(x_1) - \alpha(a)] + [\alpha(x_2) - \alpha(x_1)] + \ldots. + [\alpha(b) - \alpha(x_{n-1})]\}$$

$$= k[\alpha(b) - \alpha(a)]$$

So

$$\int_{-a}^{b} kd\alpha = \sup L(\Delta, k, \alpha) = k[\alpha(b) - \alpha(a)].$$

Also

$$U(\Delta, k, \alpha) = k[\alpha(b) - \alpha(a)]$$

and

$$\int_a^{-b} kd\alpha = \inf U(\Delta, k, \alpha) = k[\alpha(b) - \alpha(a)].$$

Thus

$$\int_a^{-b} kd\alpha = \int_{-a}^{b} kd\alpha = k[\alpha(b) - \alpha(a)]$$

and hence the result follows.

(d) Show that $f(x) = x$ is Riemann integrable in $[0, a]$.

Let Δ be a partition of $[0, a]$ into n equal subintervals.

Then

$$\Delta = \left\{0, \frac{a}{n}, \frac{2a}{n}, \ldots, \frac{na}{n} = a\right\}.$$

The length of each subinterval is $\frac{a}{n}$. The glb and lub of the gi-

function in the subintervals are $\left[\frac{(i-1)a}{n}\right]$ and $\left[\frac{ia}{n}\right]$ respective

Therefore

$$U(\Delta, f) = \frac{a^2}{n^2}(1 + 2 + \ldots + n)$$

$$= \frac{a^2}{n^2} \frac{n(n+1)}{2}$$

$$= \frac{a^2}{2}\left(1 + \frac{1}{n}\right)$$

$$L(\Delta, f) = \frac{a^2}{n^2}[0 + 1 + \ldots + (n-1)]$$

$$= \frac{a^2}{2}\left(1 - \frac{1}{n}\right).$$

Thus

$$\text{Inf } U(\Delta, f) = \sup L(\Delta, f) = \frac{a^2}{2}$$

and hence

$$\int_0^a x^2 dx = \frac{a^2}{2}.$$

6.3 PROPERTIES OF THE INTEGRAL

6.3.1. Theorem

$f \in R(\alpha)$ on $[a, b]$ iff for every $\varepsilon > 0$ there exists a partition $\Delta \in D[a, b]$ such that

$$U(\Delta, f, \alpha) - L(\Delta, f, \alpha) < \varepsilon.$$

Proof: For every $\Delta \in D[a, b]$ we have

$$L(\Delta, f, \alpha) \leq \int_{-a}^{b} f d\alpha \leq \int_{a}^{-b} f d\alpha \leq U(\Delta, f, \alpha).$$

Let $\varepsilon > 0$ be given. Suppose that there exists a $\Delta \in D[a, b]$ such that

$$U(\Delta, f, \alpha) - L(\Delta, f, \alpha) < \varepsilon.$$

Then it follows that

$$0 \leq \int_{a}^{-b} f d\alpha - \int_{-a}^{b} f d\alpha < \varepsilon.$$

But this is true for every $\varepsilon > 0$. Therefore we must have

$$\int_a^{-b} f d\alpha = \int_{-a}^{b} f d\,\alpha = \int_a^b f d\,\alpha$$

and thus $f \varepsilon R(\alpha)$.

Conversely, let $f \varepsilon R(\alpha)$. Then

$$\int_a^b f d\alpha = \int_{-a}^{b} f d\alpha = \int_a^{-b} f d\alpha.$$

But

$$\int_a^{-b} f d\alpha = \inf_{\Delta \varepsilon D[a,\, b]} U(\Delta, f, \alpha)$$

and

$$\int_{-a}^{b} f d\alpha = \sup_{\Delta \varepsilon D[a,\, b]} L(\Delta, f, \alpha).$$

Therefore for given $\varepsilon > 0$ there exist $\Delta_1, \Delta_2 \,\varepsilon\, D[a, b]$ such that

$$U(\Delta_1, f, \alpha) - \int_a^b f d\alpha < \frac{\varepsilon}{2}.$$

and

$$\int_a^b f d\alpha - L(\Delta_2, f, \alpha) < \frac{\varepsilon}{2}.$$

Let $\Delta = \Delta_1 \cup \Delta_2$. Then

$$\begin{aligned} U(\Delta, f, \alpha) \le U(\Delta_1, f, \alpha) &< \int_a^b f d\alpha + \frac{\varepsilon}{2} \\ &< L(\Delta_2, f, \alpha) + \frac{\varepsilon}{2} + \frac{\varepsilon}{2} \\ &< L(\Delta, f, \alpha) + \varepsilon \end{aligned}$$

Therefore

$$U(\Delta, f, \alpha) - L(\Delta, f, \alpha) < \varepsilon,$$

and this completes the proof.

6.3.2. Theorem

If f is continuous on $[a, b]$, then $f \varepsilon R(\alpha)$ on $[a, b]$.

Proof: Let $\varepsilon > 0$ be given. Choose $\eta > 0$ such that

$$\eta\,[\alpha(b) - \alpha(a)] < \varepsilon.$$

Assume that f is continuous on $[a, b]$. Since $[a, b]$ is compact, f is uniformly continuous. Therefore there exists a $\delta > 0$ such that

$$|f(x) - f(y)| < \eta \text{ for } |x - y| < \delta,\ x, y \,\varepsilon\, [a, b].$$

Let $\Delta \,\varepsilon\, D[a, b]$ be such that $\max \Delta x_i < \delta$.

Therefore

$$M_i - m_i \leq \eta \quad (i = 1, 2, \ldots, n)$$

Now

$$U(\Delta, f, \alpha) - L(\Delta, f, \alpha) = \sum_{i=1}^{n} (M_i - m_i)\,\Delta\alpha_i$$

$$\leq \eta \sum_{i=1}^{n} \Delta\alpha_i$$

$$= \eta\,[\alpha(b) - \alpha(a)] < \varepsilon$$

This shows that $f \varepsilon R(\alpha)$.

6.3.3. Theorem

If f is continuous on $[a, b]$, then for every $\varepsilon > 0$ there exists a $\delta > 0$ such that

$$\left| \sum_{i=1}^{n} f(t_i)\,\Delta\alpha_i - \int_a^b f\,d\alpha \right| < \varepsilon$$

for every partition $\Delta = \{x_0, x_1, \ldots, x_n\}$ of $[a, b]$ with $\max \Delta\alpha_i < \delta$, and for every choice of points $t_i \,\varepsilon\, [x_{i-1}, x_i]$.

Proof: If f is continuous on $[a, b]$, then it follows from the previous theorem that $f \varepsilon R(\alpha)$, i.e., $\int_a^b f\,d\alpha$ exists. Let $\varepsilon > 0$ be given. It clearly follows that there exists a $\delta > 0$ such that

$$\sum_{i=1}^{n} f(t_i)\, \Delta \alpha_i \text{ and } \int_a^b f d\alpha \text{ both lie between}$$

$L(\Delta, f, \alpha)$ and $U(\Delta, f, \alpha)$, for any partition $\Delta \varepsilon D[a, b]$ with max $\Delta \alpha_i < \delta$ and for all points $t_i \varepsilon [x_{i-1}, x_i]$. Therefore

$$\left| \sum_{i=1}^{n} f(t_i)\, \Delta \alpha_i - \int_a^b f d\alpha \right| < U(\Delta, f, \alpha) - L(\Delta, f, \alpha) < \varepsilon$$

for $\Delta \varepsilon D[a, b]$ with max $\Delta \alpha_i < \delta$ and for any $t_i \varepsilon [x_{i-1}, x_i]$.

In other words,

$$\lim_{\max \Delta\alpha_i \to 0} \sum_{i=1}^{n} f(t_i)\, \Delta \alpha_i = \int_a^b f d\alpha,$$

and this completes the proof.

6.3.4. Examples

(a) Let f : [-1, 1] → [0, 1] be defined by

$$f(x) = |x|.$$

Prove that f is Riemann integrable and show that

$$\int_{-1}^{1} f(x) dx = 1.$$

The function f is bounded and continuous on [-1, 1] and

$$f(x) = \begin{cases} -x, & x \le 0 \\ x, & x \ge 0. \end{cases}$$

Let Δ be a partition which divides [-1, 1] into $2n$ subintervals of equal lengths. To be precise, let

$$\Delta = \{-1 = x_0, x_1, \ldots, x_n = 0 = y_0, y_1, \ldots, y_n = 1\}.$$

Then

$$\Delta x_i = \Delta y_i = \frac{1}{n} \text{ and hence}$$

$$\lim_{n \to \infty} \Delta x_i = \lim_{n \to \infty} \Delta y_i = \lim_{n \to \infty} \frac{1}{n} = 0.$$

$$\sum_{i=1}^{n} \Delta x_i = \sum_{i=1}^{n} \Delta y_i = 1, \text{ Let } x_i = -1 + \frac{i}{n}, \; y_i = \frac{i}{n}$$

Let $t_i = x_i \in \Delta x_i,\ t_i = y_i \in \Delta y_i$. Then

$$\sum_{i=1}^{n} f(t_i)\, \Delta x_i + \sum_{i=1}^{n} f(t_i)\, \Delta y_i$$

$$= \Sigma(-x_i\, \Delta x_i) + \Sigma\, y_i\, \Delta y_i$$

$$= \Sigma\left(1 - \frac{i}{n}\right) \Delta x_i + \Sigma \frac{i}{n}\, \Delta y_i$$

$$= \Sigma\, \Delta x_i - \frac{1}{n^2} \sum_{i=1}^{n} i + \frac{1}{n^2} \sum_{i=1}^{n} i$$

$$= 1$$

Thus

$$\int_{-1}^{1} f dx = 1.$$

(b) Compute the following Riemann integral as the limit of a sum:

$$\int_{0}^{a} x^2 dx$$

Let Δ be a partition of $[0, a]$ into n equal subintervals:

$$x_0 = 0,\ x_1 = \Delta x,\ x_2 = 2\,\Delta x,\ \ldots,\ x_n = a = n\Delta x.$$

So $\Delta x = \frac{a}{n}$. Let the points t_i be chosen as the right extremities of each of the subintervals. Then

$$\sum_{i=1}^{n} f(t_i)\, \Delta x_i = x_1^2\, \Delta x + x_2^2\, \Delta x + \ldots + x_n^2\, \Delta x$$

$$= (\Delta x)^2\, \Delta x + (2\Delta x)^2\, \Delta x + \ldots + (n\Delta x)^2\, \Delta x$$

$$= (\Delta x)^3\, (1^2 + 2^2 + \ldots + n^2)$$

$$= (\Delta x)^3\, \frac{n(n+1)(2n+1)}{6}$$

$$= \frac{a^3}{n^3} \frac{n(n+1)\ (2n+1)}{6}$$

$$= \frac{a^3}{6}\left(1+\frac{1}{n}\right)\left(2+\frac{1}{n}\right)$$

$$\to \frac{a^3}{6} \cdot 2 = \frac{a^3}{3} \text{ as } \Delta x \to 0.$$

Hence $\int_0^a x^2 dx = \frac{a^3}{3}$.

(c) Evaluate $\int_a^b e^x dx$ as a limit of a sum. Let Δ be the partition of $[a, b]$ into n equal subintervals:

$$x_0 = a,\ x_1 = a + \Delta x,\ \ldots,\ x_n = a + n\Delta x, \text{ so that}$$

$\Delta x = \frac{b-a}{n}$. Take the left extreme points of each subinterval as the points t_i.

Then we have

$$\sum_{i=1}^{n} f(t_i)\,\Delta x_i = e^a\,\Delta x + e^{a+\Delta x}\,\Delta x + \ldots + e^{a+(n-1)\Delta x}\,\Delta x$$

$$= e^a\,\Delta x\,(1 + e^{\Delta x} + \ldots + e^{(n-1)\Delta x})$$

$$= e^a\,\Delta x\,\frac{e^{n\Delta x} - 1}{e^{\Delta x} - 1}$$

$$= e^a\,(e^{n\Delta x} - 1)\,\frac{\Delta x}{e^{\Delta x} - 1}$$

$$= e^a(e^{b-a} - 1)\,\frac{\Delta x}{e^{\Delta x} - 1} \quad (\text{since } n\Delta x = b - a)$$

Now $\lim\limits_{\Delta x \to 0} \frac{\Delta x}{e^{\Delta x} - 1} = \lim\limits_{\Delta x \to 0} \frac{1}{e^{\Delta x}}$ (by L'Hospital's rule)

$$= 1.$$

Therefore

$$\lim_{\Delta x \to 0} \sum_{i=1}^{n} f(t_i)\,\Delta x_i = e^a(e^{b-a} - 1) = e^b - e^a.$$

So that

$$\int_a^b e^x\,dx = e^b - e^a.$$

6.3.5. Theorem

If f is monotonic and α is continuous on $[a, b]$, then $f \varepsilon R(\alpha)$. (Of course α is also assumed to be monotonic).

Proof: Let $\varepsilon > 0$ be given. Let α be continuous and monotonically increasing on $[a, b]$. But $[a, b]$ is connected, so by the Intermediate Value Theorem α assumes every value between $\alpha(a)$ and $\alpha(b)$. Therefore for any positive integer n we can choose a partition $\Delta = x_0, x_1, \ldots, x_n$ of $[a, b]$ such that

$$\Delta\alpha_i = \frac{\alpha(b) - \alpha(a)}{n}.$$

Let m_i and M_i respectively denote the lower and upper bounds of f in $[x_{i-1}, x_i]$. Let f be monotonically increasing. Then

$$m_i = f(x_{i-1}),\ M_i = f(x_i).$$

So

$$\begin{aligned}
U(\Delta, f, \alpha) - L(\Delta, f, \alpha) &= \sum_{i=1}^{n} (M_i - m_i)\,\Delta\alpha_i \\
&= \frac{\alpha(b) - \alpha(a)}{n} \sum_{i=1}^{n} [f(x_i) - f(x_{i-1})] \\
&= \frac{\alpha(b) - \alpha(a)}{n} \cdot [f(b) - f(a)] \\
&< \varepsilon \text{ for } n > \frac{[\alpha(b) - \alpha(a)]\,[f(b) - f(a)]}{\varepsilon}
\end{aligned}$$

Thus $f \varepsilon R(\alpha)$ and this completes the proof.

6.3.6. Theorem

(a) If $f_1, f_2 \varepsilon R(\alpha)$ on $[a, b]$, then $f_1 + f_2 \varepsilon R(\alpha)$ and

$$\int_a^b (f_1 + f_2)\,d\alpha = \int_a^b f_1\,d\alpha + \int_a^b f_2\,d\alpha$$

If $f \varepsilon R(\alpha)$ on $[a, b]$ and c is any real number, then $cf \varepsilon R(\alpha)$ and

$$\int_a^b cf\, d\alpha = c \int_a^b f d\alpha$$

(b) If $f_1, f_2 \varepsilon R(\alpha)$ and $f_1 \leq f_2$ on $[a, b]$, then

$$\int_a^b f_1 d\alpha \leq \int_a^b f_2\, d\alpha$$

(c) If $f \varepsilon R(\alpha)$ on $[a, b]$ and $c \varepsilon (a, b)$, then $f \varepsilon R(\alpha)$ on $[a, c]$ and on $[c, b]$ and

$$\int_a^b f d\alpha = \int_a^c f d\alpha + \int_c^b f d\alpha$$

(d) If $f \varepsilon R(\alpha_1)$ and $f \varepsilon R(\alpha_2)$, then $f \varepsilon R(\alpha_1 + \alpha_2)$ and

$$\int_a^b f d(\alpha_1 + \alpha_2) = \int_a^b f d\alpha_1 + \int_a^b f d\alpha_2$$

If $f \varepsilon R(\alpha)$ and $c > 0$, then $f \varepsilon R(c\alpha)$ and

$$\int_a^b f d(c\alpha) = c \int_a^b f d\alpha$$

(e) If $f \varepsilon R(\alpha)$, then $|f| \varepsilon R(\alpha)$ and

$$\left| \int_a^b f d\alpha \right| \leq \int_a^b |f| d\alpha.$$

If $f \varepsilon R(\alpha)$ and if $|f(x)| \leq M$ on $[a, b]$, then

$$\left| \int_a^b f d\alpha \right| \leq M[\alpha(b) - \alpha(a)].$$

(f) If $f \varepsilon R(\alpha)$, then $f^2 \varepsilon R(\alpha)$. More generally, if $f, g \varepsilon R(\alpha)$ on $[a, b]$, then $fg \varepsilon R(\alpha)$.

Proof: (a) Let $f = f_1 + f_2$ and let Δ be any partition of $[a, b]$. Let $m_i', M_i', m_i'', M_i'', m_i, M_i$ be the bounds of f_1, f_2 and f respectively on Δx_i. Then

$$m_i' + m_i'' \le m_i \le M_i \le M_i' + M_i''.$$

Therefore

$L(\Delta, f_1, \alpha) + L(\Delta, f_2, \alpha) \le L(\Delta, f, \alpha) \le U(\Delta, f, \alpha) \le U(\Delta, f_1, \alpha) + U(\Delta, f_2, \alpha)$

Let $\varepsilon > 0$ be given. Since $f_1 \varepsilon R(\alpha)$ and $f_2 \varepsilon R(\alpha)$, there are partitions Δ_1, Δ_2 such that for $j = 1, 2$.

$$U(\Delta_j, f_j, \alpha) - L(\Delta_j, f_j, \alpha) < \frac{\varepsilon}{2}.$$

Let $\Delta = \Delta_1 \cup \Delta_2$. Then for $j = 1, 2$,

$$U(\Delta, f_j, \alpha) - L(\Delta, f_j, \alpha) < \frac{\varepsilon}{2}.$$

Therefore

$$U(\Delta, f, \alpha) - L(\Delta, f, \alpha) \le U(\Delta, f_1, \alpha) + U(\Delta, f_2, \alpha) - L(\Delta, f_1, \alpha) - L(\Delta, f_2, \alpha) < \frac{\varepsilon}{2} + \frac{\varepsilon}{2} = \varepsilon.$$

Hence $f \varepsilon R(\alpha)$.

Now since the upper integral is the infimum of upper sums, we have for $j = 1, 2$,

$$U(\Delta_j, f_j, \alpha) < \int_a^b f_j d\alpha + \frac{\varepsilon}{2},$$

also for $j = 1, 2$,

$$U(\Delta, f_j, \alpha) < \int_a^b f_j d\alpha + \frac{\varepsilon}{2}.$$

Hence

$$\int_a^b f d\alpha \le U(\Delta, f, \alpha) \le U(\Delta, f_1, \alpha) + U(\Delta, f_2, \alpha)$$

$$< \int_a^b f_1 d\alpha + \int_a^b f_2 d\alpha + \varepsilon.$$

Since $\varepsilon > 0$ is arbitrary we have

$$\int_a^b f d\alpha \le \int_a^b f_1 d\alpha + \int_a^b f_2 d\alpha.$$

If we proceed with $-f_1$ and $-f_2$ instead of f_1 and f_2 we obtain the reverse inequality and hence we have

$$\int_a^b f d\alpha = \int_a^b f_1 d\alpha + \int_a^b f_2 d\alpha.$$

The proof of the other part of (a) and that of (b) is left to the reader.

(d) Let $f \varepsilon R(\alpha_1)$ and $f \varepsilon R(\alpha_2)$. Let $\varepsilon > 0$ be given. Then there exist partitions Δ_1 and Δ_2 such that for $j = 1, 2$,

$$U(\Delta_j, f, \alpha_j) - L(\Delta_j, f, \alpha_j) < \frac{\varepsilon}{2}.$$

Let $\Delta = \Delta_1 \cup \Delta_2$. Then for $j = 1, 2$,

$$U(\Delta, f, \alpha_j) - L(\Delta, f, \alpha_j) < \frac{\varepsilon}{2}.$$

Let m_i, M_i be the bounds of f in Δx_i. Let $\alpha = \alpha_1 + \alpha_2$. Then for $j = 1, 2$,

$$\Delta \alpha_{ji} = \alpha_j(x_i) - \alpha_j(x_{i-1})$$

So

$$\begin{aligned} \Delta \alpha &= \alpha(x_i) - \alpha(x_{i-1}) \\ &= \alpha_1(x_i) + \alpha_2(x_i) - \alpha_1(x_{i-1}) - \alpha_2(x_{i-1}) \\ &= \Delta \alpha_{1i} + \Delta \alpha_{2i} \end{aligned}$$

Hence

$$U(\Delta, f, \alpha) = U(\Delta, f, \alpha_1) + U(\Delta, f, \alpha_2)$$

and so

$$\begin{aligned} U(\Delta, f, \alpha) - L(\Delta, f, \alpha) &= U(\Delta, f, \alpha_1) - L(\Delta, f, \alpha_1) \\ &\quad + U(\Delta, f, \alpha_2) - L(\Delta, f, \alpha_2) \\ &< \frac{\varepsilon}{2} + \frac{\varepsilon}{2} = \varepsilon. \end{aligned}$$

So $f \varepsilon R(\alpha)$ where $\alpha = \alpha_1 + \alpha_2$. Now

$$\begin{aligned} \int_a^b f d\alpha &= \inf U(\Delta, f, \alpha) \\ &= \inf [U(\Delta, f, \alpha_1) + U(\Delta, f, \alpha_2)] \\ &\geq \inf U(\Delta, f, \alpha_1) + \inf U(\Delta, f, \alpha_2) \end{aligned}$$

$$= \int_a^b fd\,\alpha_1 + \int_a^b fd\,\alpha_2.$$

Further $\int_a^b fd\alpha = \sup\ L(\Delta, f, \alpha)$

$$\leq \int_a^b fd\alpha_1 + \int_a^b fd\alpha_2.$$

Thus $\int_a^b fd\alpha = \int_a^b fd\alpha_1 + \int_a^b fd\alpha_2$

The proof of the other part of (d) is left to the reader as an exercise.

(e) Since $f \varepsilon R(\alpha)$, given $\varepsilon > 0$ there exists a partition Δ such that

$$U(\Delta, f, \alpha) - L(\Delta, f, \alpha) < \varepsilon..$$

Let m_i, M_i, m_i', M_i' be the bounds of f and $|f|$ respectively in Δx_i.
Then observe that

$$M_i' - m_i' \leq M_i - m_i$$

and hence

$$U(\Delta, |f|, \alpha) - L(\Delta, |f|, \alpha) \leq U(\Delta, f, \alpha) - L(\Delta, f, \alpha) < \varepsilon.$$

Hence $|f| \varepsilon R(\alpha)$.

Since $f(x) \leq |f(x)| = |f|(x)$ for all $x \varepsilon [a, b]$, it follows from (b) that

$$\int_a^b fd\,\alpha \leq \int_a^b |f|\,d\,\alpha.$$

Again, since $-f(x) \leq |f|(x)$ it follows that

$$-\int_a^b fd\,\alpha = \int_a^b (-f)d\,\alpha \leq \int_a^b |f|\,d\,\alpha.$$

Thus $\left|\int_a^b fd\alpha\right| \leq \int_a^b |f|\,d\alpha.$

(f) Since $f \varepsilon R(\alpha)$, $|f| \varepsilon R(\alpha)$. Also f is bounded, therefore $|f|$ is bounded. Thus there exists a constant M such that $|f(x)| \leq M$ for all $x \varepsilon [a, b]$. Let $\varepsilon > 0$ be given. Then there exists a partition Δ such that

$$U(\Delta, |f|, \alpha) - L(\Delta, |f|, \alpha) < \frac{\varepsilon}{2M}.$$

If m_i and M_i are bounds of f on Δx_i, then m_i^2 and M_i^2 are bounds of f^2 on Δx_i. Now

$$U(\Delta, f^2, \alpha) - L(\Delta, f^2, \alpha) = \sum_{i=1}^{n} (M_i^2 - m_i^2)\, \Delta\alpha_i$$

$$= \sum_{i=1}^{n} (M_i + m_i)(M_i - m_i)\, \Delta\alpha_i$$

$$\leq 2M \sum_{i=1}^{n} (M_i - m_i)\, \Delta\alpha_i$$

$$= 2M\, [U(\Delta, |f|, \alpha) - L(\Delta, |f|, \alpha]$$

$$< 2M \cdot \frac{\varepsilon}{2M} = \varepsilon.$$

Thus $f^2 \varepsilon R(\alpha)$.

Now assume that f, g are in $R(\alpha)$. We have

$$fg = \frac{1}{4}\, [(f+g)^2 - (f-g)^2]$$

Since $f, g \,\varepsilon\, R(\alpha)$, $f + g$, $(f+g)^2$, $(f-g)^2$ are all in $R(\alpha)$. Therefore $fg \,\varepsilon\, R(\alpha)$ and this completes the proof.

(g) Let $f \varepsilon R(\alpha)$ on $[a, b]$ and $c \,\varepsilon\, (a, b)$. Let $\varepsilon > 0$ be given. Then there exists a partition Δ such that

$$U(\Delta, f, \alpha) - L(\Delta, f, \alpha) < \varepsilon.$$

Let $\Delta^* = \Delta \cup \{c\}$. Then

$$L(\Delta, f, \alpha) \leq L(\Delta^*, f, \alpha) \leq U(\Delta^*, f, \alpha) \leq U(\Delta, f, \alpha), \text{ so that}$$

$$U(\Delta^*, f, \alpha) - L(\Delta^*, f, \alpha) \leq U(\Delta, f, \alpha) \leq L(\Delta, f, \alpha) < \varepsilon.$$

Let Δ_1, Δ_2 respectively denote the sets of points of Δ^* between $[a, c]$ and $[c, b]$. Then Δ_1, Δ_2 are partitions of $[a, c]$ and $[c, b]$ and $\Delta^* = \Delta_1 \cup \Delta_2$.

Further

$$U(\Delta^*, f, \alpha) = U(\Delta_1, f, \alpha) + U(\Delta_2, f, \alpha)$$

$$L(\Delta^*, f, \alpha) = L(\Delta, f, \alpha) + L(\Delta_2, f, \alpha)$$

So

$$[U(\Delta_1, f, \alpha) - L(\Delta_1, f, \alpha) + U(\Delta_2, f, \alpha) - L(\Delta_2, f, \alpha)]$$

$$= U(\Delta^*, f, \alpha) - L(\Delta^*, f, \alpha) < \varepsilon.$$

Therefore

$$U(\Delta_1, f, \alpha) - L(\Delta_1, f, \alpha) < \varepsilon$$

and

$$U(\Delta_2, f, \alpha) - L(\Delta_2, f, \alpha) < \varepsilon$$

and hence $f \varepsilon R(\alpha)$ on $[a, c]$ and $[c, b]$.

Now since

$$U(\Delta^*, f, \alpha) = U(\Delta_1, f, \alpha) + U(\Delta_2, f, \alpha)$$

for any partitions Δ_1, Δ_2 of $[a, c]$ and $[c, b]$ with $\Delta^* = \Delta_1 \cup \Delta_2$ and since $f \varepsilon R(\alpha)$ on $[a, c]$, $[c, b]$ and $[a, b]$ by taking the infimum of the above equation over all partitions we obtain

$$\int_a^b f d\alpha \geq \int_a^c f d\alpha + \int_c^b f d\alpha.$$

By taking $(-f)$ in place of f we obtain the reverse inequality and we therefore get

$$\int_a^b f d\alpha = \int_a^c f d\alpha + \int_c^b f d\alpha.$$

6.3.7. Theorem (First Mean Value Theorem of Integral Calculus)

Let f be continuous and α monotonically increasing on $[a, b]$. Then there exists a point $c \varepsilon [a, b]$ such that

$$\int_a^b f d\alpha = f(c)\,[\alpha(b) - \alpha(a)].$$

Proof: Since f is continuous and α is monotonic, $f \varepsilon R(\alpha)$. Since f is continuous on the compact set $[a, b]$, it is bounded. Let m and M be the infimum and supremum of f on $[a, b]$. Then

$$m[\alpha(b) - \alpha(a)] \leq \int_a^b f d\alpha \leq M[\alpha(b) - \alpha(a)].$$

So there exists a number $\mu \in [m, M]$ such that

$$\int_a^b f d\alpha = \mu [\alpha(b) - \alpha(a)].$$

Since f is continuous it assumes every value between m and M. Hence there exists a point $c \in [a, b]$ such that $f(c) = \mu$.
Therefore

$$\int_a^b f d\alpha = f(c)[\alpha(b) - \alpha(a)]$$

and this completes the proof.

6.3.8. Corollary

If f is continuous on $[a, b]$, then there exists a point $c \in [a, b]$ such that

$$\int_a^b f dx = f(c)[b - a].$$

6.4 INTEGRATION AND DIFFERENTIATION

6.4.1. Theorem

Let f be Riemann integrable on $[a, b]$ and let

$$F(x) = \int_a^x f(t)\, dt, \ x \in [a, b].$$

Then F is continuous on $[a, b]$. Further, if f is continuous at a point $x \in [a, b]$, then F is differentiable at x and

$$F'(x) = f(x).$$

Proof: Since f is integrable, it is bounded. So there exists a constant $M > 0$ such that

$$|f(x)| \leq M \text{ for all } x \in [a, b]$$

Let $x_1, x_2 \varepsilon [a, b]$ with $a \leq x_1 < x_2 \leq b$. Then

$$| F(x_2) - F(x_1) | = | \int_{x_1}^{x_2} f(t) \, dt |$$

$$\leq \int_{x_1}^{x_2} | f(t) | \, dt$$

$$\leq M (x_2 - x_1)$$

Let $\varepsilon > 0$ be given. Choose $\delta = \frac{\varepsilon}{M}$. Then

$$| F(x_2) - F(x_1) | < \varepsilon \text{ for } | x_2 - x_1 | < \delta.$$

Hence F is continuous on $[a, b]$.

Now suppose that f is continuous at $x \varepsilon [a, b]$. Then for given $\varepsilon > 0$ there exists a $\delta > 0$ such that

$$| f(y) - f(x) | < \varepsilon \text{ for } | y - x | < \delta.$$

Choose $0 < h < \delta$. Then

$$\left| \frac{F(x+h) - F(x)}{h} - f(x) \right| = \left| \frac{1}{h} \int_{x}^{x+h} f(t) \, dt - f(x) \right|$$

$$= \left| \frac{1}{h} \int_{x}^{x+h} [f(t) - f(x)] \, dt \right|$$

$$\leq \frac{1}{h} \int_{x}^{x+h} | f(t) - f(x) | \, dt$$

$$< \frac{1}{h} . \varepsilon . h = \varepsilon.$$

Thus $F'(x) = f(x)$ and this completes the proof.

6.4.2. Theorem (Fundamental Theorem of Calculus)

Let f be Riemann integrable on $[a, b]$ and let there exist a differentiable function F such that $F' = f$ on $[a, b]$. Then

$$\int_{a}^{b} f(x) \, dx = F(b) - F(a).$$

Proof: Let $\varepsilon > 0$ be given. Since f is integrable on $[a, b]$, there exists a partition Δ of $[a, b]$ such that $U(\Delta, f) - L(\Delta, f) < \varepsilon$. Let this partition Δ be given by

$$\Delta = \{ a = x_0, x_1, x_2, \ldots, x_n = b\},\ x_0 < x_1 < \ldots < x_n.$$

F is differentiable and hence we can apply Lagrange's Mean Value Theorem to obtain a point $t_i \varepsilon [x_{i-1}, x_i]$ such that for each i = 1, 2, ..., n,

$$\begin{aligned} F(x_i) - F(x_{i-1}) &= F'(t_i)\ \Delta x_i \\ &= f(t_i)\ \Delta x_i \end{aligned}$$

Therefore

$$\begin{aligned} \sum_{i=1}^{n} f(t_i)\ \Delta x_i &= \sum_{i=1}^{n} [F(x_i) - F(x_{i-1})] \\ &= F(b) - F(a). \end{aligned}$$

Now by applying Theorem 6.3.3 we obtain

$$\left| \int_a^b f(x)dx - [F(b) - F(a) \right| < \varepsilon$$

Since $\varepsilon - 0$ is arbitrary, this proves the result.

Theorem (Integration by Parts):

If u and v are differentiable functions in $[a, b]$, then

$$\int_a^b u dv = uv \Big|_a^b - \int_a^b v du.$$

Proof: We know that

$$(uv)' = u'v + uv'.$$

where u' denotes the derivative of u. Therefore

$$\int_a^b (uv)'\, dx = \int_a^b u' v dx + \int_a^b uv'\, dx$$

or, $$uv \Big|_a^b = \int_a^b v du + \int_a^b u dv$$

or, $$\int_a^b u dv = uv \Big|_a^b - \int_a^b v du.$$

Theorem (Change of variables):

Let $f(x)$ be a Riemann integrable function defined on $[a, b]$ and let $x = \varphi(t)$ where $t \varepsilon [\alpha, \beta]$. If

(i) $\varphi(\alpha) = a, \ \varphi(\beta) = b,$

(ii) $\varphi(t)$ and $\varphi'(t)$ are continuous on $[\alpha, \beta]$,

(iii) $f[\varphi(t)]$ is defined and continuous on $[\alpha, \beta]$, then

$$\int_a^b f(x)\, dx = \int_\alpha^\beta f[\varphi(t)]\, \varphi'(t)\, dt.$$

Proof: If $F(x)$ is an antiderivative of $f(x)$, then

$$\int f(x)\, dx = F(x) + C, \tag{1}$$

Hence

$$\int f[\varphi(t)]\ \varphi'(t)\, dt = F[\varphi(t)] + C, \tag{2}$$

From (1) we get

$$\int_a^b f(x)\, dx = F(x) \Big|_a^b = F(b) - F(a).$$

From (2) it follows that

$$\int_a^b f[\{\varphi(t)]\ \varphi'(t)\, dt = F[\varphi(t)]\,\Big|_\alpha^\beta$$

$$= F[\varphi(\beta)] - F[\varphi(\alpha)]$$

$$= F(b) - F(a)$$

and this completes the proof.

6.5 FUNCTIONS OF BOUNDED VARIATION

6.5.1. Definition

If f is a real function defined on $[0, 1]$, then f is said to be of bounded variation on $[0, 1]$ if

$$V(f) = \sup \sum_{i=0}^{n} | f(x_{i+1}) - f(x_i) |$$

is finite, where the supremum is taken over all partitions, $0 = x_0 < x_1 < \ldots < x_n = 1$. We write $BV[0, 1]$ to denote the set of all functions of bounded variation on $[0, 1]$.

6.5.2. Theorem

A function of bounded variation possesses a limit from the left and right at each point.

Proof: Let f be a function on $[0, 1]$ which does not have a limit from the left at some x in $[0, 1]$. We shall show that $f \notin BV[0, 1]$.

Since f does not have a limit from the left at x, for some $\varepsilon > 0$, it is true that for every $\delta > 0$ there exist s and s' in $[0, 1]$ such that $x - \delta < s < s' < x$ and $|f(s) - f(s')| \geq \varepsilon$. Thus we can choose sequences $\{s_n\}$ and $\{s_n'\}$ such that $0 < s_1 < s_1' < \ldots s_n < s_n' < x$ and $|f(s_n) - f(s_n')| \geq \varepsilon$. Now consider the partition $x_0 = 0$, $x_{2k+1} = s_k$ for $k = 0, 1, \ldots, N - 1$, $x_{2k} = s_k$ for $k = 1, 2, \ldots, N$, and $x_{2N+1} = 1$. Then

$$\sum_{k=0}^{2N} | f(x_{k+1}) | \geq \sum_{n=1}^{N} | f(s_n) - f(s_n') | \geq N\varepsilon,$$

which implies that $f \notin BV[0, 1]$. In a similar manner it can be proved that f has a limit from the right.

6.5.3. Corollary

If $f \varepsilon BV[0, 1]$, then f can have, at most, countably many discontinuities.

Proof: Observe first that f fails to be continuous at x in $[0, 1]$ if and only if $f(x) \neq f(x^+)$ or $f(x) \neq f(x^-)$. Moreover, if $x_0, x_1, \ldots, x_n$ are distinct points of $[0, 1]$, then

$$\sum_{i=0}^{N} | f(x_i) - f(x_i^+) | + \sum_{i=0}^{N} | f(x_i) - f(x_i^-) | \leq || f ||.$$

Thus for each $\varepsilon > 0$ there exists, at most, finitely many points x in $[0, 1]$ such that $| f(x) - f(x^+) | + | f(x) - f(x^-) | \geq \varepsilon$. Hence the set of discontinuities of f is at most countable.

6.5.4. Some Properties

(a) A bounded monotonic function is a function of bounded variation.

Let f be monotonically increasing on [0, 1]. Let $0 = x_0 < x_1 < \ldots < x_n = 1$. Then

$$\sum_{i=0}^{n-1} |f(x_{i+1}) - f(x_i)| = \sum_{i=0}^{n-1} \{f(x_{i+1}) - f(x_i)\}$$
$$= f(1) - f(0).$$

Thus

$$V(f) = \sup \sum_{i=0}^{n-1} |f(x_{i+1}) - f(x_i)|$$
$$= f(1) - f(0)$$

Hence a bounded monotonically increasing function is of bounded variation. The case for monotonically decreasing bounded function can be dealt in a similar way.

(b) A function of bounded variation is necessarily bounded.

Let f be of bounded variation on [0, 1]. For any $x \varepsilon [0, 1]$ consider the partition {0, x, 1} consisting of just three points, Now

$$|f(x) - f(0)| + |f(1) - f(x)| \leq V(f)$$

Thus

$$|f(x) - f(0)| \leq V(f).$$

Again
$$|f(x)| = |f(0) + f(x) - f(0)|$$
$$\leq |f(0)| + |f(x) - f(0)|$$
$$\leq f(0) + V(f)$$

Thus f is bounded on [0, 1].

(c) If the derivative f' of a function f exists and is bounded, then f is of bounded variation.

Let f' be bounded on [0, 1]. Then there exists a constant K such that

$$|f'(x)| \leq K \text{ for all } x \varepsilon [0, 1].$$

Let $P = \{0 = x_0, x_1, \ldots, x_n = 1\}$ be any partition of [0, 1]. Then

$$\sum_{i=0}^{n-1} |f(x_{i+1}) - f(x_i)| = \sum_{i=0}^{n-1} |(x_{i+1} - x_i) f'(\xi)|, \text{ for } \xi \varepsilon (0, 1), \leq K.$$

Thus $V(f)$ is finite and hence f is of bounded variation.

6.5.5. Examples

(a) Show that the following function is of bounded variation in [0, 1]:

$$f(x) = \begin{cases} x^2 \sin\left(\frac{1}{x}\right) & \text{if } x \neq 0, \\ 0 & \text{if } x = 0. \end{cases}$$

We have for $x \neq 0$,

$$f'(x) = 2x \sin\frac{1}{x} - \cos\frac{1}{x}$$

and

$$f'(0) = \lim_{x \to 0} \frac{x^2 \sin\frac{1}{x} - 0}{x}$$

$$= \lim_{x \to 0} x \sin\frac{1}{x} = 0.$$

Also $|f'(x)| \leq 3$ for $x \varepsilon [0, 1]$.

Thus f' exists and is bounded in [0, 1] and hence f is of bounded variation in [0, 1].

(b) Show that the function f defined by

$$f(x) = \begin{cases} x \sin\left(\frac{\pi}{x}\right) & \text{if } 0 < x \leq 1 \\ 0 & \text{if } \quad x = 0. \end{cases}$$

is not of bounded variation in [0, 1].

This function is continuous in [0, 1] and hence this example shows that a continuous function may not be a function of bounded variation. The proof of this is left to the reader as an exercise.

[*Hints:* Choose the partition]

$$\left\{0, \frac{2}{2n+1}, \frac{2}{2n-1}, \ldots, \frac{2}{5}, \frac{2}{3}, 1\right\}.$$

Then

$$\sum_{i=0}^{n-1} | f(x_{i+1}) - f(x_i) |$$

$$= 4\left[\frac{1}{3} + \frac{1}{5} + \frac{1}{7} + \ldots + \frac{1}{2n+1}\right].$$

Since

$$\left[\frac{1}{3}+\frac{1}{5}+\frac{1}{7}+\ldots \text{ is divergent, } V(f)=\infty\right]$$

(c) Prove that the function f defined by

$$f(x) = [x]$$

is a function of bounded variation in [0, 2]. This function is not continuous and hence this example shows that a function of bounded variation is not necessarily continuous. The proof is left to the reader.

6.5.6. Some Properties

(a) The sum of two functions of bounded variation is a function of bounded variation.

Let f and g be of bounded variation in [0, 1]. Then for any partition

$$\{0 = x_0, x_1, \ldots, x_n = 1\}$$

We have

$$\Sigma \mid (f+g)(x_{i+1}) - (f+g)(x_i) \mid$$
$$= \Sigma \mid f(x_{i+1}) + g(x_{i+1}) - f(x_i) - g(x_i) \mid$$
$$\neq \Sigma \mid f(x_{i+1}) - f(x_i) + g(x_{i+1}) - g(x_i) \mid$$
$$\leq \Sigma \mid f(x_{i+1}) - f(x_i) \mid + \Sigma \mid g(x_{i+1}) - g(x_i) \mid$$
$$\leq V(f) + V(g)$$

Thus $f+g$ is of bounded variation in [0, 1].

(b) The product of two functions of bounded variation is a function of bounded variation.

Let f and g be functions of bounded variation in [0, 1]. Then they are bounded. Thus there exists a constant K such that for all $x \varepsilon [0, 1]$,

$$| f(x) | \leq K, \ | g(x) | \leq K$$

Let $\{0 = x_0, x_1, \ldots, x_n = 1\}$
be any partition. Then

$$\Sigma \mid (fg)(x_{i+1}) - (fg)(x_i) \mid$$
$$= \Sigma \mid f(x_{i+1})\ g(x_{i+1}) - f(x_i)\ g(x_i) \mid$$
$$= \Sigma \mid f(x_{i+1}) \{g(x_{i+1}) - g(x_i)\} + g(x_i) \{f(x_{i+1}) - f(x_i)\} \mid$$

$\leq \Sigma \mid f(x_{i+1}) \mid \mid g(x_{i+1}) - g(x_i) \mid + \mid g(x_i) \mid \mid f(x_{i+1}) - f(x_i) \mid$
$\leq K V(g) + KV(f).$

Thus fg is of bounded variation.

We shall close this section with a very important theorem known as the Jordan Theorem which gives a characterisation of a function of bounded variation.

6.5.7. Jordan Theorem

A function of bounded variation can be expressed as the difference of two monotonically increasing functions. To be more precise, if f is a function of bounded variation on [0, 1], then there exist monotonically increasing functions f_1 and f_2 on [0, 1] such that for $x \varepsilon$ [0, 1],

6.5.8.

$$f(x) = f_1(x) - f_2(x)$$

6.5.9

$$V(f) = f_1(x) + f_2(x).$$

Proof: Define f_1 and f_2 by

$$f_1 = \frac{1}{2}[V(f) + f],$$

$$f_2 = \frac{1}{2}[V(f) - f].$$

Thus 6.5.8 and 6.5.9 hold.

It is now enough to show that f_1 and f_2 are monotonically increasing. For that, let $x_2 > x_1$. Then

$$f_1(x_2) - f_1(x_1)$$

$$= \frac{1}{2}[V(f(x_2)) + f(x_2)] - \frac{1}{2}[V(f(x_1) + f(x_1)]$$

$$= \frac{1}{2}[V(f(x_2)) - V(f(x_1))] + \frac{1}{2}[f(x_2) - f(x_1)]$$

$$= \frac{1}{2}[V(f) + \{f(x_2) - f(x_1)\}]$$

Since $V(f) \geq \mid f(x_2) - f(x_1) \mid$ we get

$$f_1(x_2) - f_1(x_1) \geq 0$$

$$\text{i.e., } f_1(x_2) \geq f_1(x_1),$$

so that f_1 is monotonically increasing.
Similarly it can be shown that

$$f_2(x_2) \geq f_2(x_1)$$

and this completes the proof of the theorem.

6.6 IMPROPER INTEGRAL

We have discussed the concept of the Riemann integral of a bounded function f defined on a (bounded) closed interval $[a, b]$ in the real line R. In this section we shall attach a meaning to the integral of a function f if

(i) either the interval is of the form $[a, \infty)$ or $(\infty, b]$ where $a, b \in R$ or if

(ii) the value of the function tends to infinity as x tends to some point $c \in [a, b]$. Such integrals will be called *improper integrals.*

Definition: Suppose f is defined and bounded in $[a, \infty]$ where $a \in R$. Then the integral of f on $[a, \infty]$ is defined to be

$$\lim_{y \to \infty} \int_a^y f(x)dx,$$

if it exists, and is written as

$$\int_a^\infty f(x)dx.$$

If the above limit exists, then we say that the improper integral

$$\int_a^\infty f(x)dx$$

converges. Otherwise the integral is said to be divergent. Similarly if f is bounded on $[-\infty, b]$, and if

$$\lim_{y \to -\infty} \int_y^b f(x)dx$$

exists, then we say that the improper integral

$$\int_{-\infty}^{b} f(x)dx$$

is convergent.

If f is defined and bounded on R, then the improper integral

$$\int_{-\infty}^{\infty} f(x)\,dx = \int_{-\infty}^{a} f(x)\,dx + \int_{a}^{\infty} f(x)\,dx,$$

if both the integrals on the right converge for some $a \in R$.
The improper integrals defined above, that is,

$$\int_{a}^{\infty} f(x)dx, \quad \int_{-\infty}^{b} f(x)dx, \text{ and } \int_{-\infty}^{\infty} f(x)dx$$

are called the *improper integrals of the first type.*

Definition: Let f be defined on $[a, b]$, but $\to \infty$ as $x \to a$. Let $h > 0$. Then the improper integral

$$\int_{a}^{b} f(x)dx$$

is defined to be

$$\lim_{h \to 0} \int_{a+h}^{b} f(x)dx,$$

if the limit exists and in this case the integral is said to be convergent. Similarly if $f(x) \to \infty$ as $x \to b$, then for $h > 0$,

$$\int_{a}^{b} f(x)dx = \lim_{h \to 0} \int_{a}^{b-h} f(x)dx$$

if the limit exists.

If $f(x) \to \infty$ as $x \to c \in (a,b)$, then we examine the convergence of the improper integrals

$$\int_{a}^{c} f(x)\,dx \text{ and } \int_{c}^{b} f(x)dx.$$

If both these integrals converge, then we define

$$\int_a^b f(x)dx = \int_a^c f(x)dx + \int_c^b f(x)dx.$$

Such integrals are called *improper integrals of the second type.*

We state the following two theorems which deal with the tests for convergence of improper integrals.

Theorem: Let f and g be two real-valued functions which are defined and positive on $[a, \infty]$ and let

$$\lim_{x \to \infty} \frac{f(x)}{g(x)} = l$$

where l is a nonzero real number. Then the integrals

$$\int_a^\infty f(x)dx \text{ and } \int_a^\infty g(x)dx$$

either both converge or both diverge.

Theorem: Let f and g be two functions which are bounded and positive on $[a + h, b]$, $h > 0$, and let "a" be the only point of infinite discontinuity. Let

$$\lim_{x \to a+} \frac{f(x)}{g(x)} = l,$$

where l is a nonzero real number. Then the integrals

$$\int_a^b f(x)dx \text{ and } \int_a^b g(x)dx$$

either both converge or both diverge at a.

By using the concept of improper integral we have a test for convergence of an infinite series. This is known as the Integral Test for Convergence and is applied as follows:

Theorem: Let the terms of the series Σu_n be positive and nonincreasing, that is,

$$u_1 \geq u_2 \geq \ldots \geq u_n \geq \ldots$$

and let f be a continuous real-valued nonincreasing function such that $f(n) = u_n$ for all $n = 1, 2, \ldots.$. Then the series Σu_n converges if and only if the improper integral.

$$\int_1^{\infty} f(x)dx$$

converges.

We shall not prove this theorem here, but we shall show that the convergence of the series $\Sigma \frac{1}{n^p}$ can be very easily proved with the help of the integral test.

Example: Show that $\Sigma \frac{1}{n^p}$ converges for $p > 1$ and diverges for $0 \leq p \leq 1$.

To prove this take

$$f(x) = \frac{1}{x^p}.$$

This function f satisfies all the conditions of the theorem. Now for $p \neq 1$,

$$\int_1^N \frac{dx}{x^p} = \frac{1}{1-p} x^{1-p} \Big|_1^N = \frac{1}{1-p} (N^{1-p} - 1).$$

If $p > 1$,

$$\int_1^{\infty} \frac{dx}{x^p} = \lim_{n \to \infty} \int_1^N \frac{dx}{x^p} = \frac{1}{1-p}.$$

Thus the improper integral $\int_1^{\infty} \frac{dx}{x^p}$ converges and hence the series $\Sigma \frac{1}{n^p}$ converges.

If $p < 1$, $\int_a^{\infty} \frac{dx}{x^p}$ diverges and thus $\Sigma \frac{1}{n^p}$ diverges.

If $p = 1$, then

$$\int_1^N \frac{dx}{x^p} = \int_1^N \frac{dx}{x} = \ln x \Big|_1^N = \ln N$$

$\to \infty$ as $N \to \infty$.

Therefore $\Sigma \frac{1}{n}$ diverges.

We now discuss two important examples of improper integrals. One of them is called the gamma function and the other the beta function in the regions where these integrals converge. We first discuss the gamma function. Consider the integral

$$\int_0^\infty e^{-x}\, x^{a-1}\, dx$$

where a is a constant. If $a \geq 1$, the integrand is continuous at $x = 0$ and hence the integral is improper of the first type. But if $a < 1$, the integrand $\to \infty$ as $x \to 0$ and hence the above integral is improper at both the limits. Therefore we have to consider the convergence of the integrals.

$$\int_0^1 e^{-x}\, x^{a-1}\, dx \text{ and } \int_1^\infty e^{-x}\, x^{a-1}\, dx$$

at 0 and ∞ respectively.

For convergence at 0, put

$$f(x) = e^{-x}\, x^{a-1} = \frac{e^{-x}}{x^{1-a}}$$

and define

$$F(x) = \frac{1}{x^{1-a}}.$$

Now

$$\lim_{x \to 0} \frac{f(x)}{F(x)} = 1$$

and $\int_0^1 F(x)\, dx = \int_0^1 \frac{1}{x^{1-a}}\, dx.$

is convergent at 0 iff $a > 0$.

For convergence at ∞. observe that

$$e^x > x^{a+1}$$

for every constant a. Thus

$$e^{-x} > x^{-a-1}$$

and hence

$$e^{-x}\, x^{a-1} < x^{a-1}\, x^{-a-1} = \frac{1}{x^2}.$$

Therefore,

$\int_1^x e^{-x} x^{a-1} dx$ is bounded above, and thus

$$\int_1^\infty e^{-x} x^{a-1} dx$$

is convergent at ∞ for every value of a. Thus it follows that

$$\int_0^\infty e^{-x} x^{a-1}\, dx$$

is convergent for $a > 0$ and therefore defines a function of a. This function is called the gamma function and is denoted by

$$\Gamma(a) = \int_0^\infty e^{-x} x^{a-1}\, dx, \ (a > 0).$$

Observe that

$$\Gamma(1) = \int_0^\infty e^{-x}\, dx = -e^{-x} \Big|_0^\infty = 1.$$

We have the following result, called the *reduction formula* for $\Gamma(a)$.

(i) $\Gamma(a + 1) = a\,\Gamma(a)$.

(ii) If $a = n$ (an integer) then $\Gamma(n + 1) = n\,!$

Proof: We have

(i) $\Gamma\,(a + 1) = \int_0^\infty e^{-x}\, x^a\, dx$

$$= -x^{-a}\, e^{-x} \Big|_0^\infty + a \int_0^\infty e^{-x}\, x^a\, dx$$

$$= a \int_0^\infty e^{-x}\, x^{a-1}\, dx = a\ \Gamma\ (a).$$

(ii) $\Gamma(n+1) = n\Gamma(n) = n(n-1)\Gamma(n-1)$

$$= n(n-1)(n-2)\ \Gamma\ (n-2)$$

$$= \dots \quad \dots. \quad \dots$$

$$= n(n-1) \ \dots \ 1\ \Gamma(1)$$

$$= n\,!$$

Beta function: Consider the integral

$$\int_0^1 x^{m-1}(1-x)^{n-1}\,dx.$$

Observe that

(i) for $m \geq 1$, the integrand is continuous at $x = 0$ and for $m < 1$, the integrand $\to \infty$ as $x \to 0$.

(ii) for $n \geq 1$, the integrand is continuous at $x = 1$ and for $n < 1$, the integrand $\to \infty$ as $x \to 1$.

(iii) the integrand is continuous for every value of x other than 0 and 1.

Thus the integral is proper for $m \geq 1$, $n \geq 1$.

We now assume that $m < 1$ or $n < 1$ and examine the convergence of the integrals

$$\int_0^{1/2} x^{m-1}\ (1-x)^{n-1}\ dx \text{ and } \int_{1/2}^1 x^{m-1}\ (1-x)^{n-1}\ dx$$

at 0 and 1 respectively.

Convergence at 0: Put

$$f(x) = x^{m-1}(1-x)^{n-1} = \frac{(1-x)^{n-1}}{x^{1-m}}$$

and define

$$F(x) = \frac{1}{x^{1-m}}.$$

Now

$$\frac{f(x)}{F(x)} \to 1 \text{ as } x \to 0.$$

Since $\int_0^{1/2} F(x)\,dx$

converges at 0 iff $1 - m < 1$ i.e. iff $m > 0$, it follows that

$$\int_0^{1/2} x^{m-1} (1-x)^{n-1}\, dx.$$

converges at 0 iff $m > 0$.

Convergence at 1 : Put

$$f(x) = x^{m-1} (1-x)^{n-1} = \frac{x^{m-1}}{(1-x)^{1-n}}$$

Define

$$F(x) = \frac{1}{(1-x)^{1-n}}.$$

It can be shown that the integral

$$\int_{1/2}^{1} x^{m-1} (1-x)^{n-1}\, dx$$

converges at 1 iff $n > 0$.

Thus

$$\int_0^1 x^{m-1} (1-x)^{n-1}\, dx$$

is meaningful if $m > 0$ and $n > 0$ and in this case it is called beta function and is written

$$\beta(m, n) = \int_0^1 x^{m-1} (1-x)^{n-1}\, dx, \ (m > 0,\ n > 0).$$

Observe that if we put $x = \sin^2 \theta$ in the above integral we have

$$\beta(m, n) = 2 \int_0^{\pi/2} \sin^{2m-1} \theta \cos^{2n-1} \theta \, d\theta.$$

There is an important relation between the beta and gamma function and this is given by

$$\beta(m, n) = \frac{\Gamma(m)\, \Gamma(n)}{\Gamma(m+n)}, \ (m > 0,\ n > 0).$$

The proof of this result uses the concept of double integral and hence it is beyond the scope of this book. Nevertheless, we prove the following result by using the above relation.

$$\Gamma\left(\frac{1}{2}\right) = \sqrt{\pi}.$$

By taking $m = n = \frac{1}{2}$ in the relation

$$\frac{\Gamma(m)\,\Gamma(n)}{\Gamma(m+n)} = \beta(m, n)$$

We obtain

$$\frac{[\Gamma(\frac{1}{2})]^2}{\Gamma(1)} = \beta\left(\frac{1}{2}, \frac{1}{2}\right)$$

$$= 2\int_0^{1/2} \cos^0\theta \,\sin^0\theta \, d\theta$$

$$= \pi.$$

But $\Gamma(1) = 1$. Hence $\Gamma\left(\frac{1}{2}\right) = \sqrt{\pi}$.

EXERCISES

1. Answer Yes or No

(i) All bounded functions are Riemann integrable.

(ii) Partition Δ of the interval $[a, b]$ should always have unequal subintervals.

(iii) Every discontinuous function is Riemann integrable.

(iv) For a constant function, the upper sum and the lower sum are equal.

(v) The integer function $[x]$ is Riemann integrable.

(vi) A function $f(x)$ is Riemann-Stieltjes integrable with respect to $\alpha(x)$ if $\alpha(x)$ is bounded and continuous.

(vii) A function is not Riemann-Stieltjes integrable if it is unbounded at lower, upper or both the end points of the interval.

2. State which of the following are correct:

(i) $U(\Delta_1, f) - U(\Delta_2, f) \geq 0$ for $\Delta_1 \subset \Delta_2$

(ii) $\int_{-a}^{b} f d\alpha = \int_{-a}^{c} f d\alpha + \int_{-c}^{b} f d\alpha$ for $a < c < b$

(iii) $\int_{-a}^{b} f d\alpha \neq \int_{a}^{b} f d\alpha$ where

$$f(x) = \begin{cases} 0 & \text{for } x \in N, \\ 1 & \text{for } x \notin N. \end{cases}$$

(iv) $\int_{a}^{b} f d\alpha = \int_{a}^{b} f \alpha \, dx$

(v) $\lim_{\varepsilon \to 0} \int_{a+\varepsilon}^{b} f dx = \lim_{\varepsilon \to 0} \int_{a}^{b-\varepsilon} f dx = \int_{a}^{b} f dx$

(vi) $L(\Delta, f) + U(\Delta^*, f) > L(\Delta^*, f) + U(\Delta, f)$

where Δ^* is refinement of Δ.

(vii) $L(\Delta, f_1 f_2) = L(\Delta, f_1) \; L(\Delta, f_2)$

(viii) $m(b-a) \le \int_{a}^{b} f dx \le M(b-a)$

where $m = \inf_{x \in [a, b]} f(x)$ and $M = \sup_{x \in [a, b]} f(x)$

(ix) $\frac{d}{dx} \int_{a}^{x} f(x) dx = f(x)$

3. Give an example in each of the following

(i) $f(x)$ and $\alpha(x)$, where $f(x)$ is Riemann integrable but not Riemann-Stieltjes integrable with respect to α.

(ii) $f(x)$ where $f(x)$ is not Riemann integrable but $|f(x)|$ is.

(iii) A discontinuous function which is Riemann-Stieltjes integrable with respect to a suitable α.

(iv) A discontinuous function which is not Riemann integrable.

4. Compute each of the following Riemann integrals as a limit of a sum and show that

(i) $\int_{-1}^{1} |x|\, dx = 1,$

(ii) $\int_{a}^{b} c\, xdx = c\,\frac{b^2 - a^2}{2},$

(iii) $\int_{a}^{b} \frac{dx}{x} = ln\ \frac{b}{a}$ where $0 < a < b,$

(iv) $\int_{a}^{b} \sqrt{x}dx = \frac{2}{3}(b^{3/2} - a^{3/2})$

5. Show that

$$\int_{0}^{3} s d([|x|] - x) = \frac{3}{2}$$

6. Define f on [0, 1] by

$$f(x) = \begin{cases} 0, & x \text{ irrational or zero} \\ \frac{1}{n} & \text{when } x = \frac{m}{n} \text{ (a rational number with least postitive integers } m \text{ and } n) \end{cases}$$

Show that f is Riemann integrable and $\int_{0}^{1} fdx = 0$

7. Define f on [-1, 1] by

$$f(x) = \begin{cases} k, & x \neq 0 \text{ (}k\text{ is a positive constant)} \\ 0, & x = 0 \end{cases}$$

Show that f is Riemann integrable and $\int_{-1}^{1} fdx = 2k$

8. Show by examples that if $|f|$ is Riemann integrable, then f need not be so.

[*Hint:* take $f(x) = \begin{cases} 1, & x \text{ rational} \\ -1, & x \text{ irrational} \end{cases}$

As another example, take

$$f(x) = \begin{cases} x, & x \text{ rational} \\ -x, & x \text{ irrational}] \end{cases}$$

9. Show that f defined by

$$f(x) = \begin{cases} \frac{1}{2^n}, & \frac{1}{2^{n+1}} < x \le \frac{1}{2^n}, \quad n = 0, 1, \ldots \\ 0, & x = 0 \end{cases}$$

is Riemann integrable in [0, 1] although it has infinitely manydiscontinuities.

[*Hint:* f is bounded and monotonically increasing and hence is integrable]

10. If the two functions $f(x)$ and $g(x)$ are Riemann integrable on [a, b], then show that

$$\max_{x \,\varepsilon\, [a,\, b]} [f(x), g(x)]$$

and

$$\min_{x \,\varepsilon\, [a,\, b]} [f(x), g(x)]$$

are also Riemann integrable on $[a, b]$.

11. Prove that if $f(x) \geq 0$ on $[a, b]$ then $\int_a^b f \, d\alpha = 0$ implies that either $a = b$ or $f(x) = 0$ on $[a, b]$.

12. If $f(x)$ is Riemann integrable and $f(x) = g(x)$ on $[a, b]$, except for a countable set of points, then

$$\int_a^b f(x) dx = \int_a^b g(x) dx$$

7 Sequences and Series of Functions

7.1 UNIFORM CONVERGENCE

7.1.1. Definition

Let X be any set and let $\{f_n\}$ be a sequence of real functions each defined on X. Then for each $x \varepsilon X$, $\{f_n(x)\}$ is a sequence of real numbers.

The sequence $\{f_n\}$ of functions is said to converge pointwise to a function f if the sequence $\{f_n(x)\}$ converges to $f(x)$ for each $x \varepsilon X$, that is, if for each $x \varepsilon X$,

$$\lim_{n\to\infty} f_n(x) = f(x)$$

Similarly, if the series $\sum_n f_n(x)$ converges to $f(x)$ for each $x \varepsilon X$, then the series $\sum_n f_n$ is said to converge pointwise to the function f and this is written as

$$\sum_n f_n(x) = f(x), \ (x \varepsilon X)$$

Certain natural and pertinent questions arise in this connection. These concern whether or not the limits defined above for a pointwise convergent sequence of functions preserve certain important and commonly used properties of functions. For example, whether the limit of a pointwise convergent sequence of—

(i) continuous functions is continuous;
(ii) differentiable functions is differentiable;
(iii) integrable functions is integrable; and so on.

The answer is "no" in general, which will be clear from the following examples.

7.1.2. Examples

(a) The sequence $\{f_n\}$ where

$$f_n(x) = \begin{cases} 1 - nx^2 & \text{for } x \,\varepsilon \left[0, \dfrac{1}{\sqrt{n}}\right] \\ 0 & \text{for } x \,\varepsilon \left[\dfrac{1}{\sqrt{n}}, 1\right] \end{cases}$$

converges. Each term of the sequence is continuous, but the limit function is not so.

If $f(x)$ is the limit of this sequence, then

$$f(x) = \lim_{n \to \infty} f_n(x) = 0 \text{ for } x \neq 0.$$

However $f_n(0) = 1$ for each n.

But $f(0) = 0$ and hence

$$\lim_{x \to 0} \lim_{n \to \infty} f_n(x) \neq \lim_{n \to \infty} \lim_{x \to 0} f_n(x)$$

$f(x)$ is obviously discontinuous at $x = 0$.

(b) For the sequence $\{f_n\}$ where

$$f_n(x) = nx(1 - x^2)^n, \; x \,\varepsilon\, [0, 1], \; n \,\varepsilon\, N,$$

the limit of the integrals is not equal to the integral of the limit.

$$\text{For } 0 < x \leq 1, \quad \lim_{n \to \infty} f_n(x) = 0$$

At $x = 0$, $f_n(0) = 0$ for each n. So

$$\lim_{n \to \infty} f_n(0) = 0.$$

Thus the limit function

$$f(x) = \lim_{n \to \infty} f_n(x) = 0 \text{ for } x \,\varepsilon\, [0, 1].$$

Therefore

$$\int_0^1 f(x)dx = 0.$$

But

$$\int_0^1 f_n(x)dx = \int_0^1 nx(1-x^2)^n \, dx = \frac{n}{2n+2} = \frac{1}{2+\frac{2}{n}}$$

Therefore

$$\lim_{n\to\infty} \int_0^1 f_n(x)dx = \lim_{n\to\infty} \frac{1}{2+\frac{2}{n}} = \frac{1}{2},$$

and so

$$\lim_{n\to\infty} \int_0^1 f_n(x)dx \neq \int_0^1 f(x)dx.$$

(c) For the sequence $\{f_n\}$ where

$$f_n(x) = \frac{\sin nx}{\sqrt{n}}, \; x \in R$$

the limit of the derivative is not equal to the derivative of the limit. We have for $x \in R$,

$$f(x) = \lim_{n\to\infty} f_n(x) = 0$$

So $f'(x) = 0$ and hence $f'(0) = 0$.
But $f_n'(x) = \sqrt{n} \text{ Cos } nx$,
and therefore

$$f_n'(0) = \sqrt{n} \to \infty \text{ as } n \to \infty,$$

Thus at $x = 0$, the sequence $\{f_n'(x)\}$ diverges whereas the derivative of the limit of $\{f_n(x)\}$ is 0.

(d) The series Σf_n where $f_n(x) = (x^n - x^{n+1})$ converges to $f(x)$ for $x \in [0, 1]$, each term $f_n(x)$ of the series is continuous but $f(x)$ is discontinuous at $x = 1$.
Here $\lim_{x\to 1} f_n(x) = 0$ for all $n \in N$

Also

$$f(x) = \begin{cases} 1 & \text{for } x \neq 1 \\ 0 & \text{for } x = 1 \end{cases}$$

as the given series is geometric.

Thus $f(x)$ is discontinuous.

The above examples motivate us to define a new type of convergence for sequences of functions, which should be naturally stronger than pointwise convergence and which will preserve the important properties of functions, such as continuity etc. This is called *uniform convergence* and its formal definition is given below.

7.1.3. Definition

A sequence $\{f_n\}$ of real functions, each defined on a set X, is said to converge uniformly on X to a function f if, for every $\varepsilon > 0$, there exists an integer $n_0 \in N$ such that

7.1.4.

$$|f_n(x) - f(x)| < \varepsilon \text{ for } n \geq n_0, \ x \in X.$$

7.1.5. Remark

We now explicitly point out the essential difference between the two concepts: pointwise and uniform convergence.

If $\{f_n\}$ converges pointwise to f on X, then for every $\varepsilon > 0$ and for every $x \in X$ there exists $n_0 \in N$. depending on ε and x, such that 7.1.4 holds for $n \geq n_0$. For different $x \in X$, we may get different n_0 which satisfies 7.1.4 and it may not be possible to select one n_0 for which 7.1.4 holds for every $x \in X$. But if $\{f_n\}$ is uniformly convergent to f on X, then for every $\varepsilon > 0$ there exists n_0 depending only on ε , such that 7.1.4 holds for all $x \in X$.

Thus every uniformly convergent sequence of functions is obviously pointwise convergent. The converse is not always true and this will be clear from the examples given below. We shall first present an alternative characterisation of uniform convergence before giving the examples.

7.1.6. Theorem (Cauchy Criterion)

A sequence $\{f_n\}$ of real functions defined on a set X converges uniformly on X iff for every $\varepsilon > 0$, there exists $n_0 \in N$ (depending only on ε) such that

$$| f_n(x) - f_m(x) | < \varepsilon \text{ for } n, m \geq n_0, \ x \in X.$$

Proof: Suppose $\{f_n\}$ converges uniformly to f on X. Let $\varepsilon > 0$ be given. Then there exists $n_0 \varepsilon N$ such that

$$| f_n(x) - f(x) | < \frac{\varepsilon}{2} \text{ for } n \geq n_0,\ x \varepsilon X.$$

Let n, $m \geq n_0$ and $x \varepsilon X$. Then

$$|f_n(x) - f_m(x)| \leq |f_n(x) - f(x)| + |f_m(x) - f(x)| < \frac{\varepsilon}{2} + \frac{\varepsilon}{2} = \varepsilon.$$

Conversely, suppose that the given condition holds. Therefore $\{f_n(x)\}$ is a Cauchy sequence in R for each $x \varepsilon X$. Since R is complete, it follows that $\{f_n(x)\}$ converges to some value $f(x)$ for each $x \varepsilon X$ and hence $\{f_n\}$ converges to f pointwise.

We need only to show that the convergence is uniform.

To show this let $\varepsilon > 0$ be given. Then by the hypothesis, there exists $n_0 \varepsilon N$ (depending only on ε) such that

$$|f_n(x) - f_m(x)| < \varepsilon \text{ for } n, m \geq n_0,\ x \varepsilon X.$$

Let n be fixed and let $m \to \infty$. Then we have

$$|f_n(x) - f(x)| < \varepsilon \text{ for } n \geq n_0,\ x \varepsilon X.$$

and this completes the proof.

7.1.7. Theorem

A series of real functions Σf_n, each defined on a set X converges uniformly on X iff for every $\varepsilon > 0$ there exists $n_0 \varepsilon N$ (depending only on ε) such that

$$| f_{n+1}(x) + f_{n+2}(x) + \ldots + f_{n+m}(x) | < \varepsilon \text{ for } n \geq n_0,\ m \geq 1,\ x \varepsilon X.$$

The proof of this is left to the reader as an exercise.

7.1.8. Examples

(a) Show that the sequence $\{f_n\}$ given by

$$f_n(x) = \tan^{-1} nx,\ x \geq 0$$

is uniformly convergent in any interval $[a, b]$, $a > 0$, and is not uniformly convergent (but pointwise convergent) in $[0, b]$.

In this case the pointwise limit f of $\{f_n\}$ is given by

$$f(x) = \lim_{n\to\infty} f_n(x) = \begin{cases} \frac{\pi}{2}, & x > 0 \\ 0 & x = 0 \end{cases}$$

Let $\varepsilon > 0$ be given. Then for $x > 0$,

$$| f_n(x) - f(x) | = | \tan^{-1} nx - \frac{\pi}{2} | < \varepsilon$$

if $\cot^{-1} nx < \varepsilon$ i.e., if $n > \cot \frac{\varepsilon}{x}$.

Now $\cot \frac{\varepsilon}{x}$ decreases with x and its maximum value is $\cot \frac{\varepsilon}{x}$. Choose an integer n_0 such that $n_0 \geq \cot \frac{\varepsilon}{a}$. Then for all $x \varepsilon [a, b]$,

$$| f_n(x) - f(x) | < \varepsilon \text{ for } n \geq n_0$$

Hence $\{f_n\}$ converge uniformly in $[a, b]$. However, $\cot \frac{\varepsilon}{x} \to \infty$ as $x \to 0$, so that in this case no n_0 exists for which

$$| f_n(x) - f(x) | < \varepsilon \text{ for } n \geq n_0$$

Thus sequence is not uniformly convergent in $[0, b]$.

(b) The sequence $\{f_n\}$ given by

$$f_n(x) = \frac{nx}{1 + n^2 x^2}$$

is not uniformly convergent (but pointwise convergent) on any interval $[a, b]$ containing 0.

The sequence converges pointwise to f where

$$f(x) = 0 \text{ for all } x \varepsilon R.$$

If $\{f_n\}$ converges uniformly in any interval $[a, b]$, then gives $\varepsilon > 0$ there exists an integer n_0 such that for all $x \varepsilon [a, b]$,

$$\left| \frac{nx}{1 + n^2 x^2} \right| < \varepsilon \text{ for } n \geq n_0.$$

Take $\varepsilon < \frac{1}{2}$. Choose an integer $m > n_0$ such that $\frac{1}{m} \varepsilon [a, b]$. If we take $n = m$ and $x = \frac{1}{m}$, then

$$\frac{nx}{1 + n^2 x^2} = \frac{1}{2} > \varepsilon$$

Thus the sequence is not uniformly convergent in an interval which contains the point $\frac{1}{m}$. But $\frac{1}{m} \to 0$ as $m \to \infty$ and hence $\{f_n\}$ is not uniformly convergent in an interval containing 0.

(c) The series Σf_n where

$$S_n(x) = f_1(x) + f_2(x) \ldots + f_n(x) = n\,x\,e^{-nx^2}$$

is pointwise but not uniformly convergent on any interval $[0, k]$, $k > 0$.

The series is pointwise convergent and the pointwise sum is given by

$$S(x) = \lim_{n \to \infty} S_n(x) = 0 \text{ for all } x \geq 0.$$

If the series converges uniformly on $[0, k]$, then for given $\varepsilon > 0$ there exists an integer n_0 such that for all $x \geq 0$

$$|S_n(x) - S(x)| = nxe^{-nx^2} < \varepsilon \text{ for } n \geq n_0.$$

Let n_1 be an integer greater than n_0 and $e^2\,\varepsilon^2$. Then for $x = \frac{1}{\sqrt{n_1}}$ and $n = n_1$ we have

$$\frac{\sqrt{n_1}}{e} < \varepsilon \Rightarrow n_1 < e^2\,\varepsilon^2$$

and this gives a contradiction.

7.1.9. Theorem

Let $\{f_n\}$ be a pointwise convergent sequence of real functions each defined on a set X such that

$$\lim_{n \to \infty} f_n(x) = f(x),\ x \,\varepsilon\, X,$$

and let

$$M_n = \sup_{x \varepsilon X} |f_n(x) - f(x)|$$

Then $\{f_n\}$ converges to f uniformly on X iff

$$M_n \to 0 \text{ as } n \to \infty.$$

Proof: Let $\{f_n\}$ converge to f uniformly on X. Hence for a given $\varepsilon > 0$ there exists $n_0 \,\varepsilon\, N$ such that

$$|f_n(x) - f(x)| < \varepsilon \text{ for } n \geq n_0,\ x \,\varepsilon\, X.$$

This implies that

$$M_n = \sup_{x \varepsilon x} | f_n(x) - f(x) | \leq \varepsilon, \text{ for } n \geq n_0$$

and hence $M_n \to 0$ as $n \to \infty$.

Conversely, let $M_n \to 0$ as $n \to \infty$, Then, for a given $\varepsilon > 0$ there exists $n_0 \varepsilon N$ such that

$$M_n = \sup_{x \varepsilon X} | f_n(x) - f(x) | < \varepsilon \text{ for } n \geq n_0$$

Hence

$$| f_n(x) - f(x) | < \varepsilon \text{ for } n \geq n_0,\ x \varepsilon X$$

and thus $\{f_n\}$ converges to f uniformly.

7.2 WEIERSTRASS'S M-TEST

7.2.1. Theorem (Weierstrass's M-Test)

A series Σf_n of real functions converges uniformly (and absolutely) on X if there exists a convergent series ΣM_n of positive numbers such that

$$|f_n(x)| \leq M_n \text{ for all } n \varepsilon N \text{ and all } x \varepsilon X.$$

Proof: Let $\varepsilon > 0$ be given. Since ΣM_n converges, there exists a positive integer n_0 such that

$$| M_{n+1} + M_{n+2} + \ldots + M_{n+m} | < \varepsilon \text{ for } n \geq n_0,\ m \geq 1$$

Let $x \varepsilon X$, $n \geq n_0$ and $m \geq 1$. Then

$$| f_{n+1}(x) + f_{n+2}(x) + \ldots + f_{n+m}(x) |$$

$$\leq | f_{n+1}(x) + | f_{n+2}(x) | + \ldots + | f_{n+m}(x) |$$

$$\leq M_{n+1} + M_{n+2} \ldots + M_{n+m}$$

$$< \varepsilon.$$

Thus it follows that Σf_n converges uniformly and absolutely on X and this completes the proof.

7.2.2. Examples

(a) Show that the sequence $\{f_n\}$ where

$$f_n(x) = n x e^{-nx^2},\ x \geq 0.$$

is not uniformly convergent on $[0, a]$ where $a > 0$.

In this case, for each $x \geq 0$,

$$f(x) = \lim_{n \to \infty} f_n(x) = 0.$$

f attains its maximum at $x = \frac{1}{\sqrt{2n}}$ and the maximum value $= \sqrt{\frac{n}{2e}}$.

But

$$M_n = \sup_{x \varepsilon [0,a]} | f_n(x) - f(x) |$$

$$= \sup_{x \varepsilon [0,a]} n x e^{-nx^2}$$

$$= \sqrt{\frac{n}{2e}} \quad \to \infty \text{ as } n \to \infty.$$

Thus $M_n \nrightarrow 0$ as $n \to \infty$ and hence $\{f_n\}$ is not uniformly convergent in $[0, a]$.

(b) Show that the sequence $\{f_n\}$ where

$$f_n(x) = \frac{nx}{1 + n^2x^2}$$

is not uniformly convergent in any interval containing 0.

In this case

$$\lim_{n \to \infty} f_n(x) = 0 \text{ for all } x.$$

f_n attains the maximum at $x = \frac{1}{n}$, maximum value $= \frac{1}{2}$, again $\frac{1}{n} \to 0$ as $n \to \infty$. Consider the interval $[a, b]$ such that it contains 0. Now

$$M_n = \sup_{x \varepsilon [a,b]} |f_n(x) - f(x)|$$

$$= \sup_{x \varepsilon [a,b]} \left| \frac{nx}{1 + n^2x^2} \right|$$

$$= \frac{1}{2} \nrightarrow 0 \text{ as } n \to \infty.$$

Hence $\{f_n\}$ is not uniformly convergent in $[a, b]$.

(c) Show that $\{f_n\}$ where

$$f_n(x) = \frac{x}{1 + nx^2}, \; x \,\varepsilon\, R$$

converges uniformly on any closed interval.

In this case

$$f(x) = \lim_{n \to \infty} f_n(x) = 0 \text{ for all } x \varepsilon R.$$

$f(x)$ attains maximum value $\frac{1}{2\sqrt{n}}$ at $x = \frac{1}{\sqrt{n}}$. Thus

$$M_n = \sup \mid f_n(x) - f(x) \mid = \sup \mid \frac{x}{1 + nx^2} \mid$$

$$= \frac{1}{2\sqrt{n}} \to 0 \text{ as } n \to \infty.$$

Therefore the sequence converges uniformly.

7.3 UNIFORM CONVERGENCE AND CONTINUITY

7.3.1. Theorem

The limit, f, of a uniformly convergent sequence $\{f_n\}$ of continuous real functions, each defined on $[a, b]$, is continuous, that is, for $c \varepsilon [a, b]$,

$$\lim_{x \to c}\left(\lim_{n \to \infty} f_n(x)\right) = \lim_{n \to \infty}\left(\lim_{x \to c} f_n(x)\right).$$

Proof: Since $\{f_n\}$ converges uniformly to f, given $\varepsilon > 0$ there exists $n_0 \varepsilon N$, depending only on ε, such that

$$\mid f_n(x) - f(x) \mid < \frac{\varepsilon}{3} \text{ for } x \varepsilon (a,b)$$

Choose any point $c \varepsilon [a, b]$. Since f_n is continuous at c for each $n \varepsilon N$, there exists a $\delta > 0$ such that

$$|f_n(x) - f_n(c)| < \frac{\varepsilon}{3} \text{ for } x \varepsilon (c - \delta, c + \delta)$$

Now for $x \varepsilon (c - \delta, c + \delta)$ and $n \geq n_0$ we have

$$\begin{aligned} \mid f(x) - f(c) \mid &\leq \mid f(x) - f_n(x) \mid + \mid f_n(x) - f_n(c) \mid \\ &\quad + |f_n(c) - f(c)| \\ &< \frac{\varepsilon}{3} + \frac{\varepsilon}{3} + \frac{\varepsilon}{3} \\ &= \varepsilon. \end{aligned}$$

This completes the proof.

7.4 TERM BY TERM INTEGRATION

7.4.1. Theorem

If f_n is Riemann integrable for each $n \varepsilon N$ and $\{f_n\}$ converges uniformly to f on $[a, b]$, then f is Riemann integrable on $[a, b]$ and

$$\int_a^b f dx = \lim_{n \to \infty} \int_a^b f_n dx$$

Proof: Let $\varepsilon > 0$ be given. Choose $\delta > 0$ such that

$$\delta [b - a] < \frac{\varepsilon}{3}.$$

Since $\{f_n\}$ converges to f uniformly on $[a, b]$, there exists $n_0 \varepsilon N$ such that for all $x \varepsilon [a,b]$,

$$| f_{n_0}(x) - f(x) | < \delta.$$

i.e., $f_{n_0}(x) - \delta < f(x) < f_{n_0}(x) + \delta$.

Since f_n is Riemann integrable, there exists a partition Δ of $[a, b]$ such that

$$U(\Delta, f_n) - L(\Delta, f_n) < \frac{\varepsilon}{3}.$$

Since $f(x) < f_{n_0}(x) + \delta$,

$$U(\Delta, f) < U(\Delta, f_{n_0}) + \frac{\varepsilon}{3}$$

and since $f(x) > f_{n_0}(x) - \delta$, we have

$$L(\Delta, f) > L(\Delta, f_{n_0}) - \frac{\varepsilon}{3}$$

Therefore

$$U(\Delta, f) - L(\Delta, f)$$

$$< U(\Delta, f_{n_0}) - L(\Delta, f_{n_0}) + \frac{\varepsilon}{3} + \frac{\varepsilon}{3}$$

$$< \frac{\varepsilon}{3} + \frac{\varepsilon}{3} + \frac{\varepsilon}{3} = \varepsilon.$$

This shows that $f(x)$ is Riemann integrable.

Given that $\{f_n\}$ converges to f uniformly on $[a, b]$. Therefore there exists $n_1 \varepsilon N$ such that for $n \geq n_1$ and $x \varepsilon [a, b]$,

$$| f_n(x) - f(x) | < \frac{\varepsilon}{b-a}$$

Therefore for $n \geq n_1$,

$$\begin{aligned} \left| \int_a^b f dx - \int_a^b f_n dx \right| &= \left| \int_a^b (f - f_n)\, dx \right| \\ &\leq \int_a^b | f(x) - f_n(x) |\, dx \\ &< \frac{\varepsilon}{b-a} [b-a] \\ &= \varepsilon \end{aligned}$$

and this completes the proof.

7.4.2. Corollary

If f_n is Riemann integrable on $[a, b]$ and if

$$f(x) = \sum_{n=1}^{\infty} f_n(x)$$

uniformly on $[a, b]$, then f is Riemann integrable on $[a, b]$ and

$$\int_a^b f dx = \sum_{n=1}^{\infty} \int_a^b f_n dx.$$

7.4.3. Corollary

(a) If each $f_n \varepsilon R(\alpha)$ and if $\{f_n\}$ converges to f uniformly on $[a, b]$, then $f \varepsilon R(\alpha)$ and

$$\int_a^b f d\alpha = \lim_{n \to \infty} \int_a^b f_n\, d\alpha$$

(b) If $f_n \varepsilon R(\alpha)$ and if

$$f(x) = \sum_{n=1}^{\infty} f_n(x)$$

uniformly on $[a, b]$, then $f \varepsilon R(\alpha)$ and

$$\int_a^b f d\alpha = \sum_{n=1}^{\infty} \int_a^b f_n d\alpha.$$

7.5 TERM BY TERM DIFFERENTIATION

7.5.1. Theorem

If $\{f_n\}$ is a sequence of differentiable real functions each defined on $[a, b]$ such that

(i) $\{f_n(c)\}$ converges at some point $c \varepsilon [a, b]$,

(ii) f_n' converges uniformly on $[a, b]$.

then $\{f_n\}$ converges uniformly on $[a, b]$ and for all $x \varepsilon [a, b]$,

$$f'(x) = \lim_{n \to \infty} f_n'(x).$$

Proof: Let $\varepsilon > 0$ be given. Since $\{f_n(c)\}$ converges and f' converges uniformly on $[a, b]$, there exists $n_0 \varepsilon N$ such that for $n, m \geq n_0$,

$$| f_n(c) - f_m(c) | < \frac{\varepsilon}{2}$$

and

$$| f_n'(x) - f_m'(x) | < \frac{\varepsilon}{2(b-a)}, \; x \varepsilon [a, b].$$

We can apply Lagrange's Mean Value Theorem to the function $f_n - f_m$ in any interval $[x, t] \subset [a, b]$. Therefore there exists a point $\xi \varepsilon (x, t)$ such that for $n, m \geq n_0$,

7.5.2.

$$| [f_n(x) - f_m(x)] - [f_n(t) - f_m(t)] |$$

$$\leq | x - t | [f_n'(\xi) - f_m'(\xi)]$$

$$< | x - t | \frac{\varepsilon}{2(b-a)}$$

$$< (b-a) \cdot \frac{\varepsilon}{2(b-a)} = \frac{\varepsilon}{2},$$

since $|x - t| < (b - a)$.

Now for $x \varepsilon [a, b]$ and $n, m \geq n_0$,

$$|f_n(x) - f_m(x)| \leq |[f_n(x) - f_m(x)] - [f_n(c) - f_m(c)]|$$
$$+ |f_n(c) - f_m(c)|$$
$$< \frac{\varepsilon}{2} + \frac{\varepsilon}{2} = \varepsilon.$$

Therefore $\{f_n\}$ converges uniformly on $[a, b]$. Let

$$\lim_{n \to \infty} f_n(x) = f(x),\ x \in [a, b].$$

Let $x \in [a, b]$ be chosen arbitrarily and for $t \in [a, b]$ with $t \neq x$ let

7.5.3.

$$F_n(t) = \frac{f_n(t) - f_n(x)}{t - x},\ F(t) = \frac{f(t) - f(x)}{t - x}.$$

Now

$$\lim_{t \to x} F_n(t) = f_n'(x)$$

and it follows from 7.5.2 that

$$|F_n(t) - F_m(t)| < \frac{\varepsilon}{2(b - a)},\ n, m \geq n_o.$$

Therefore $\{F_n\}$ converges uniformly for $t \neq x$, and since $\{f_n\}$ converges to f uniformly it follows from 7.5.3 that

$$\lim_{n \to \infty} F_n(t) = F(t).$$

Now

$$f'(x) = \lim_{t \to x} F(t) = \lim_{t \to x} \left(\lim_{n \to \infty} F_n(t) \right)$$
$$= \lim_{n \to \infty} \left(\lim_{t \to x} F_n(t) \right)$$
$$= \lim_{n \to \infty} f_n'(x)$$

and this completes the proof.

7.6 POWER SERIES

A power series is a formal series of the form

$$a_0 + a_1 x + a_2 x^2 + \ldots + a_n x^n + \ldots$$

where $a_0, a_1, a_2, \ldots, a_n, \ldots$ are constants and they are called the coefficients of the series.

The convergence of a power series depends on x and the domain of convergence is always some interval, which, in a particular case, can be a single point.

We have

7.6.1. Theorem (Abel)

If a power series converges for some non-zero value x_0 of x, then it converges absolutely for all values of x for which

$$|x| < |x_0|.$$

If the series diverges for some value x_0, then it diverges for every x for which

$$|x| > |x_0|$$

Proof: By the hypothesis, the series

$$a_0 + a_1 x_0 + a_2 x_0^2 + \ldots + a_n x_0^n + \ldots$$

converges. Hence $a_n x_0^n \to 0$ as $n \to \infty$. In other words, there exists a positive number M such that all the terms of the series are less than M in absolute value.

Rewrite the series in the form

$$a_0 + a_1 x_0 \left(\frac{x}{x_0}\right) + a_2 x_0^2 \left(\frac{x}{x_0}\right)^2 + \ldots + a_n x_0^n \left(\frac{x}{x_0}\right)^n + \ldots.$$

The series in absolute values of its terms is given by

$$|a_0| + |a_1 x_0| \left|\frac{x}{x_0}\right| + |a_2 x_0^2| \left|\frac{x}{x_0}\right|^2 + \ldots + |a_n x_0^n| \left|\frac{x}{x_0}\right|^n + \ldots$$

The terms of this series are less than the corresponding terms of the series

$$M + M\left|\frac{x}{x_0}\right| + M\left|\frac{x}{x_0}\right|^2 + \ldots + M\left|\frac{x}{x_0}\right|^n + \ldots$$

For $|x| < |x_0|$, this series is a geometric series of common ratio $\left|\frac{x}{x_0}\right| < 1$, and hence converges. Therefore the given power series converges absolutely for $|x| < |x_0|$. The case of divergence can be proved in a similar way.

7.6.2. Radius of convergence

We shall now give a method for determining the radius of convergence of a power series.

Let the given power series be

$$a_0 + a_1 x + a_2 x^2 + \ldots + a_n x^n + \ldots$$

Consider the following series which is in absolute values of its terms:

$$| a_0 | + | a_1 | \, | x | + | a_2 | \, | x |^2 + \ldots + |a_n| \, |x|^n + \ldots$$

Let R denote the radius of convergence of this series.

Then

$$R = \lim_{n \to \infty} | a_n |^{-\frac{1}{n}}.$$

or

$$R = \lim_{n \to \infty} \left| \frac{a_n}{a_{n+1}} \right|.$$

Now $\lim\limits_{n \to \infty} \left| \dfrac{a_{n+1} x^{n+1}}{a_n x^n} \right| = \lim\limits_{n \to \infty} \left| \dfrac{a_{n+1}}{a_n} \right| | x | = L | x |$, say.

Then by the D'Alembert ratio test, the series converges if $L | x | < 1$, i.e., if $| x | < \frac{1}{L}$ and diverges if $L | x | > 1$, i.e., if $| x | > \frac{1}{L}$. Hence

$$R = \frac{1}{L} = \lim_{n \to \infty} \left| \frac{a_n}{a^{n+1}} \right|.$$

Similarly if we apply Cauchy's root test we have

$$\lim_{n \to \infty} | a_n x^n |^{\frac{1}{n}} = \lim_{n \to \infty} | a_n |^{\frac{1}{n}} | x | = L | x |, \text{ say.}$$

Then the series converges if $L | x | < 1$, i.e., if $| x | < \frac{1}{L}$ and diverges if

$L | x | > 1$, i.e., if $| x | > \frac{1}{L}$. Hence

$$R = \frac{1}{L} = \lim_{n \to \infty} | a_n |^{-\frac{1}{n}}$$

7.6.3. Example

(a) Determine the radius and interval of convergence of the power series

$$1 + x + x^2 + \ldots + x^n + \ldots.$$

Now

$$\lim_{n \to \infty} \left| \frac{x^{n+1}}{x^n} \right| = |x|.$$

Hence the series converges for $|x| < 1$ and diverges for $|x| > 1$. Thus the radius of convergence is 1 and the interval of convergence is $(-1, 1)$.

(b) We have observed in Chapter 5 that Maclaurin's series for e^x is given by

$$e^x = 1 + x + \frac{x^2}{2!} + \frac{x^3}{3!} + \ldots.. + \frac{x^n}{n!} + \ldots.$$

In this case for all $x \varepsilon R$,

$$\lim_{n \to \infty} \left| \frac{x^{n+1}}{(n+1)!} \cdot \frac{n!}{x^n} \right| = \lim_{n \to \infty} \frac{|x|}{(n+1)} = 0$$

Hence the radius of convergence $R = \infty$.

(c) $\text{Cos } x = 1 - \frac{x^2}{2!} + \frac{x^4}{4!} - .. + (-1)^n \frac{x^{2n}}{(2n)!} - \ldots.$

Radius of convergence $R = \infty$.

(d) $\text{Sin } x = x - \frac{x^3}{3!} + \frac{x^5}{5!} - .. + (-1)^{n+1} \frac{x^{2n-1}}{(2n-1)!} \ldots.$

Radius of convergence $R = \infty$.

(e) We have for $|x| < 1$,

$$\log (1 + x) = x - \frac{x^2}{2} + \frac{x^3}{3} - \ldots.$$

$$\lim_{n \to \infty} \left| -\frac{x^{n+1}}{n+1} \cdot \frac{n}{x^n} \right| = \lim_{n \to \infty} \left| \frac{n}{n+1} x \right|$$

$$= \lim_{n \to \infty} \frac{1}{1 + \frac{1}{n}} |x| = |x|$$

Hence radius of convergence = 1.

(f) Observe that the radius of convergence for the binomial series $(1 + x)^m$, m not being a positive integer, is 1.

7.7 WEIERSTRASS APPROXIMATION THEOREM

Finally in the following sub-section we shall present a proof of the famous Weierstrass Theorem in its simplest form. It deals with the approximation by polynomials of continuous real functions defined on the compact set [0, 1].

7.7.1. Theorem (Weierstrass Approximation Theorem)

Let f be a continuous real function defined on the compact set [0, 1] and let $\varepsilon > 0$ be given. Then there exists a polynomial P with real coefficients [i.e., $P(x) = a_0 + a_1 x + \ldots a_n x^n$, $a_0, a_1, \ldots, a_n \in R$] such that

$$|f(x) - P(x)| < \varepsilon \text{ for all } x \in [0, 1].$$

Proof: First observe that if n is a positive integer and k an integer such that $0 \le k \le n$, then the binomial coefficient $\binom{n}{k}$ is defined by

$$\binom{n}{k} = \frac{n!}{k!\,(n-k)!}$$

Note that for each n, the polynomial B_n defined by

$$B_n(x) = \sum_{k=0}^{n} \binom{n}{k} x^k (1-x)^{n-k} f\left(\frac{k}{n}\right)$$

is called a Bernstein polynomial associated with the function f.

We shall first establish several identities which will be used to complete the proof.

Recall that the following identity is a special case of the binomial theorem:

$$\sum_{k=0}^{n} \binom{n}{k} x^k (1-x)^{n-k} = [x + (1-x)]^n = 1 \ldots \qquad \textbf{(7.7.2)}$$

By differentiating 7.7.2 with respect to x we get,

$$\sum_{k=0}^{n} \binom{n}{k} [k\, x^{k-1} (1-x)^{n-k} - (n-k)\, x^{k} (1-x)^{n-k-1}]$$

$$= \sum_{k=0}^{n} \binom{n}{k} x^{k-1} (1-x)^{n-k-1} (k - nx) = 0.$$

By multiplying $x(1-x)$ throughout we get

$$\sum_{k=0}^{n} \binom{n}{k} x^{k} (1-x)^{n-k} (k - nx) = 0 \ldots$$

(7.7.3)

On differentiating 7.7.3 with respect to x and considering $x^{k}(1-x)^{n-k}$ as one of the two factors in applying the product rule, we get

$$\sum_{k=0}^{n} \binom{n}{k} [-nx^{k} (1-x)^{n-k} + x^{k-1} (1-x)^{n-k-1} (k - nx)^{2}] \quad .. \textbf{(7.7.4)}$$

From 7.7.2 and 7.7.4 we get

$$\sum_{k=0}^{n} \binom{n}{k} x^{k-1} (1-x)^{n-k-1} (k - nx)^{2} = n.$$

On multiplying $x(1-x)$ throughout, we get

$$\sum_{k=0}^{n} \binom{n}{k} x^{k} (1-x)^{n-k} (k - nx)^{2} = nx(1-x)$$

On dividing both sides by n^2 we have

$$\sum_{k=0}^{n} \binom{n}{k} x^{k} (1-x)^{n-k} \left(x - \frac{k}{n} \right)^{2} = \frac{x(1-x)}{n} \quad .. \qquad \textbf{(7.7.5)}$$

By using 7.7.2, we get

$$f(x) - B_n(x) = \sum_{k=0}^{n} \binom{n}{k} x^{k} (1-x)^{n-k} \left[f(x) - f\left(\frac{k}{n}\right) \right],$$

so that

$$|f(x) - B_n(x)| \le \sum_{k=0}^{n} \binom{n}{k} x^{k} (1-x)^{n-k} \left| f(x) - f\left(\frac{k}{n}\right) \right| \qquad \textbf{(7.7.6)}$$

Since [0, 1] is compact and f is continuous, f is uniformly continuous. Hence we can find a $\delta > 0$ such that

$$|f(x) - f\left(\frac{k}{n}\right)| < \frac{\varepsilon}{2} \text{ for } |x - \frac{k}{n}| < \delta.$$

Let

$$\sum_{k=0}^{n} \binom{n}{k} x^k (1-x)^{n-k} \, | f(x) - f\left(\frac{k}{n}\right) | = \Sigma_1 + \Sigma_2$$

where Σ_1 is the sum of those terms for which $|x - \frac{k}{n}| < \delta$ and Σ_2 is the sum of the remaining terms. Observe that $\Sigma_1 < \frac{\varepsilon}{2}$. We shall now show that if n is sufficiently large, Σ_2 can be made less than $\frac{\varepsilon}{2}$ independently of x.

Since f is continuous and [0, 1] is compact, f is bounded. Hence there exists a positive real number K such that $|f(x)| \le K$ for all $x \,\varepsilon\, [0,1]$. From this it follows that

$$\Sigma_2 \le 2K \, \Sigma \binom{n}{k} x^k (1-x)^{n-k}.$$

Let the sum on the right, which is taken over all k such that $|x - \frac{k}{n}| \ge \delta$, be denoted by Σ_3. It is now enough to show that if n is sufficiently large, Σ_3 can be made less than $\frac{\varepsilon}{4K}$ independently of x. From 7.7.5 it follows that

$$\delta^2 \Sigma_3 \le \frac{x(1-x)}{n}$$

or

$$\Sigma_3 \le \frac{x(1-x)}{n\delta^2}$$

The maximum value of $x(1-x)$ on [0, 1] being $\frac{1}{4}$ we have

$$\Sigma_3 \le \frac{1}{4n\delta^2}$$

Choose $n > \frac{K}{\delta^2_{\varepsilon}}$. Then $\Sigma_3 < \frac{\varepsilon}{4K}$, so that $\Sigma_2 < \frac{\varepsilon}{2}$.

Therefore for $n > \frac{K}{\delta_\varepsilon^2}$,

$$|f(x) - B_n(x)| < \varepsilon \text{ for all } x \in [0, 1].$$

Thus $B_n(x)$ is uniformly close to $f(x)$ for all sufficiently large n and this completes the proof of the theorem.

EXERCISES

1. Answer Yes or No:

(i) For a sequence of real-valued functions pointwise convergence is the same as the convergence of a sequence of real numbers.

(ii) The subsequence of a uniformly convergent sequence is uniformly convergent.

(iii) Weierstrass's M-test is not applicable to the series of functions which are discontinuous at a finite number of points in the interval.

(iv) If a term by term integration of a given sequence is convergent then the given sequence is uniformly convergent.

(v) The sequence of the derivative of a uniformly convergent sequence is uniformly convergent.

2. State which of the following are correct:

(i) $\{x - [x]\}$, $x \in (n - 1, n)$ is uniformly convergent for $x \in R$.

(ii) Uniform convergence of $\{f_n(x)\}$ implies uniform convergence of $\Sigma f_n(x)$.

(iii) If $\{\alpha_n\}$ is uniformly convergent and $f(x)$ is bounded then

$$\left\{\int_a^b f\, d\alpha_n\right\}$$

is uniformly convergent.

3. Show that if $\{f_n\}$ and $\{g_n\}$ converge uniformly on a set E, then $\{f_n + g_n\}$ converges uniformly.

4. Show that $\{(\sin x)^{\frac{1}{n}}\}$ converges but not uniformly on $[0, \pi]$.

5. Show that $\left\{\left(\frac{\sin x}{x}\right)^{\frac{1}{n}}\right\}$ converges but not uniformly on $[0, \pi]$.

6. Show that the sequence $\{f_n\}$ defined by

$$f_n(x) = e^{-nx}$$

converges pointwise but not uniformly in $(0, \infty)$ whereas it is uniformly convergent in $[k, \infty]$ for any $k > 0$.

7. Let $f_n(x) = x^n$. Show that $\{f_n\}$ is uniformly convergent in $[0, k]$ for $0 < k < 1$. It is not uniformly convergent (but is pointwise convergent) in $[0, 1]$.

8. Show that $\{f_n\}$ where $f_n(x) = \frac{n}{x+n}$ is uniformly convergent in $[0, k]$ for all $k \in R$ but is not uniformly convergent in $(0, \infty)$.

9. If $f_n(x) = n x e^{-nx^2}$, then show that $\{f_n\}$ is pointwise convergent but not uniformly convergent in $[0, \infty]$.

10. Show that the series $\Sigma f_n(x)$ where

$$f_n(x) = \frac{x^2}{(1+x^2)^n}, \; x \in R$$

converges, each term of the series is continuous, but the sum function is not so.

11. Apply Weierstrass's M-test to show that the series

$$\Sigma \frac{\sin nx}{n^p} \text{ and } \Sigma \frac{\cos nx}{n^p}$$

are uniformly convergent for $p > 1$.

12. Discuss pointwise convergence and uniform convergence of the series

$$\Sigma \left(\frac{x}{3}\right)^n$$

13. Prove that the sequence

$$\{(1-x)x^n\}$$

converges uniformly on $[0, 1]$.

14. Prove that the series

$$\Sigma \frac{\sin nx}{n}$$

is not uniformly convergent on any interval that includes $x = 0$.

15. If $\{f_n\}$ converges uniformly on R to a continuous function f then prove that

$$\lim_{n \to \infty} f_n\left(x + \frac{1}{n}\right) = f,\ x \,\varepsilon\, R$$

16. If the series $\Sigma\, a_n$ converges and $f(x) = \sum_{n=0}^{\infty} a_n x^n$, $x \,\varepsilon\, (-1, 1)$ prove that f is continuous on $(-1, 1)$.

17. If $\Sigma |a_n| < M$, M is large, prove that

$$\int_0^1 (\Sigma\, a_n x^n)\, dx = \Sigma \frac{a_n}{n+1}$$

8 Lebesgue Measure and Integration

In Chapter 6 we discussed the concept of Riemann Integral. We observed that the mapping f defined on a closed interval $[a, b]$ by

$$f(x) = \begin{cases} 0 & \text{if } x \text{ is rational,} \\ 1 & \text{if } x \text{ is irrational,} \end{cases}$$

is not integrable in the Riemann sense. Therefore it was natural and desirable to expect that there should be an extension to the concept of Riemann integral. The new definition should be such that

(i) if a function is integrable in the Riemann sense it should also be integrable in the sense of the new definition and the values of the integrals in both the cases should be same; and

(ii) there should be some function which must be integrable in the sense of this new definition, but not integrable in the Riemann sense.

Keeping this in mind several mathematicians, towards the end of nineteenth century, proposed various schemes. The most useful and widely accepted was the definition proposed by the famous French mathematician Lebesgue at the beginning of the twentieth century. This concept was later termed the *Lebesgue integral.*

The purpose of this last chapter is to provide a brief introduction to the Lebesgue integral. In order to define this we need to introduce the concepts called *measure, measurable sets*, and *measurable functions*. We shall first discuss the concepts of measures of sets and measurable sets, and then introduce measurable functions and study their properties. Subsequently, we discuss the concept of Lebesgue integral and functions which are integrable in the Lebesgue sense.

It will be soon clear to the reader that the concept of measure is a generalisation of the length of an interval. We define measure for certain subsets of R, called measurable sets. By definition the measure of any bounded interval is its length. To be more precise, let $I_{a,b}$ be any one of the intervals (a, b), $[a, b)$, $(a, b]$ or $[a, b]$ where $a, b \in R$ and $a \leq b$. Then the measure of $I_{a,b}$ is its length, which is $(b - a)$. This being so, there should be no difficulty in comprehending the concepts of measures and measurable sets.

It is worth mentioning here that the famous British mathematician, J.E. Littlewood, made these three interesting remarks while teaching Measure theory to his students:

(i) A measurable set is 'nearly' a finite union of intervals.

(ii) A measurable function is 'nearly' a continuous function.

(iii) A convergent sequence of measurable functions is 'nearly' uniformly convergent.

Thus, according to Professor Littlewood, one who is familiar with intervals and continuous functions has no difficulty in understanding measurable sets and measurable functions. The full implications of the Professor's remarks and the sense in which he used the word 'nearly' will become clear when we discuss the basic concepts of measurable sets and functions.

8.1 INTERVALS AND SETS ON R

Let $I_{a,b}$ be a bounded interval on real line with end points a and b. In other words, $I_{a,b}$ may be any one of the intervals (a, b), $[a, b)$, $(a, b]$ or $[a, b]$. If we represent the length of an interval by l, then we have

$$l(I_{a,b}) = b - a \qquad \text{... (1)}$$

Now we define a class of sub-intervals of $I_{a,b}$ and denote it by $\{I_{a_i,b_i}\}$ $(i = 1, 2, 3, \ldots, n)$. Let these sub-intervals be pairwise disjoint. Then

$$l\left\{\bigcup_{i=1}^{n} I_{a_i,b_i}\right\} = \sum_{i=1}^{n} (b_i - a_i) \qquad \text{... (2)}$$

Also

$$l\{I_{a,b}\} \geq l\left\{\bigcup_{i=1}^{n} I_{a_i,b_i}\right\} \qquad \text{... (3)}$$

as each I_{a_i,b_i} $(i = 1, 2, \ldots, n)$ is contained in $I_{a,b}$ and all are pairwise disjoint.

We can extend the finite class of subinterval $\{I_{a_i, b_i}\}$ $(i = 1, 2, \ldots, n)$ to countable. Result (3) holds for this class also, i.e.

$$l\{I_{a,b}\} \geq l\{\bigcup_{i=1}^{\infty} I_{a_i, b_i}\}$$

as each I_{a_i, b_i} $(i = 1, 2, 3, \ldots)$ is contained in $I_{a,b}$ and are pairwise disjoint.

Now we consider open sets on R. We know that if $\{I_n\}$ $(n = 1, 2, 3, \ldots)$ is a collection of open intervals on R then the set $\bigcup_{n=1}^{\infty} I_n$ is an open set in R. Similarly, we can write the converse in the following form:

Every open subset G of R can be expressed as a countable union of pairwise disjoint open intervals. For example let G be an open subset of R, then we can write

$$G = \bigcup_{i=1}^{n} I_{a_i, b_i} \tag{4}$$

where n may be a positive integer or infinite and $\{I_{a_i, b_i}\}$, $(i = 1, 2, \ldots, n)$ are pairwise disjoint open intervals. Thus we have $l\{G\} = l\left\{\bigcup_{i=1}^{n} I_{a_i, b_i}\right\}$

or

$$l\{G\} = \sum_{i=1}^{n} l\{I_{a_i, b_i}\} \tag{5}$$

as I_{a_i, b_i} are pairwise disjoint.

For any collection $\{I_{a_i, b_i}\}$ of open intervals we have

$$l\{\bigcup_{i=1}^{n} I_{a_i, b_i}\} \leq \sum_{i=1}^{n} l\{I_{a_i, b_i}\} \tag{6}$$

n is a positive integer or infinite.

Now if we have two open subsets G_1 and G_2 of R such that $G_1 \subset G_2$, then it is easy to prove that

$$l\{G_1\} \leq l\{G_2\} \tag{7}$$

Next, we consider the length of a closed bounded subset F of R. The length of F will be taken to be zero when F is either empty or has only one element. Otherwise we consider, the smallest closed interval $I_{a,b}$

which contains F. In this case the set $I_{a,b} - F$ is open. Then the length $l\{F\}$ is defined as

$$l\{F\} = (b-a) - l\{I_{a,b} - F\} \tag{8}$$

If $I_{a,b}$ is an open interval which contains F, then

$$l\{I_{a,b} - F\} = l\{I_{a,b}\} - l\{F\} \tag{9}$$

The above discussion leads us to the following simple results:

(i) Let F_1 and F_2 be two bounded closed sets such that

$$F_1 \subset F_2, \text{ then } l\{F_1\} \le l\{F_2\}$$

(ii) Let F and G be bounded closed and bounded open sets respectively such that $F \subset G$, then

$$l\{F\} \le l\{G\} \tag{10}$$

(iii) If F^c is the complement of F, then

$$l\{F\} = (b-a) - l\{F^c\} \tag{11}$$

(iv) If a bounded open set G is the union of a countably many pairwise disjoint open sets G_i, then

$$l\{G\} = \sum_{i=1}^{\infty} l\{G_i\}$$

Finally, we state the following two theorems which will be used while discussing measure of sets.

8.1.1. Theorem

Let G be a bounded open set and $\{F_i\}$ the collection of all closed sets contained in G, then

$$l\{G\} = \operatorname*{Sup}_i l\{F_i\} \tag{12}$$

8.1.2. Theorem

Let $\{G_i\}$ be the class of all bounded sets containing a closed set F, then

$$l\{F\} = \inf_i l\{G_i\} \tag{13}$$

8.2 MEASURE OF SETS ON R

As indicated earlier, 'measure' is a generalisation of length and is assigned to certain sets on the real line to be called measurable sets. Essentially, measure is a non-negative real number and for a bounded interval its measure and length are one and the same. Measure of two disjoint sets is the sum of the measures of the sets. In fact, measure is an additive function on the class of measurable sets. However, to define the measure of a set properly, we have to define the inner and outer measures of that set.

We consider a set $E \subset I_{a,b}$. The outer measure of E, denoted by $m_0(E)$, is defined as

$$m_0(E) = \inf_i l\{G_i\} \tag{14}$$

where $\{G_i\}$ is the collection of all open sets which contain E.

The *inner measure* of E, denoted by $m_c(E)$, is defined as

$$m_c(E) = \sup_i l\{F_i\} \tag{15}$$

Where $\{F_i\}$ is the collection of all closed sets contained in E.

In other words, if G and F are open and closed sets respectively such that G, contains E, and F is contained in E, then

$$m_0(E) \leq l\{G\} \text{ and } l\{F\} \leq m_c(E) \tag{16}$$

A set E is said to be *measurable* if its inner and outer measures are equal, i.e., if

$$m_c(E) = m_0(E) \tag{17}$$

This common value is called the *measure* of the set E. We denote the measure of E by $m(E)$. This means that if E is measurable, then

$$m_c(E) = m_0(E) = m(E) \tag{18}$$

$$\text{Thus we can write } m(E) = \inf l\{G\} \tag{19}$$

provided E is measurable and the infimum is taken over all open sets G which contain E. As G is an open set it can be expressed as a countable union of pairwise disjoint open intervals on R. Therefore the length of G is an additive function and is equal to the measure of E. Thus, if E_1 and E_2 are two measurable disjoint sets, then

$$m(E_1 \cup E_2) = m(E_1) + m(E_2) \tag{20}$$

Similarly if $E_1, E_2, E_3, \ldots \ldots$ are pairwise disjoint measurable sets, then

$$m(E_1 \cup E_2 \cup E_3 \ldots) = m(E_1) + m(E_2) + m(E_3) + \ldots \tag{21}$$

These remarks give an idea about the measure of a bounded set on real line. To provide a better insight into 'measure' (including inner and outer measure) we present below some basic properties.

8.2.1. Theorem

Let E be a bounded set on R. Then

$$m_c(E) \leq m_0(E)$$

Proof: We consider a class of open sets $\{G_i\}$ such that each G_i contains E and a class of closed sets $\{F_j\}$ such that each F_j is contained in E, i.e.

$$F_j \subset E \subset G_i \Rightarrow F_j \subset G_i$$

This implies that

$$l\{F_j\} \leq l\{G_i\}$$

This relation is true for all j and all i. Hence

$$\sup_j l\{F_j\} \leq \inf_i l\{G_i\}$$

i.e.,
$$m_c(E) \leq m_0(E) \tag{22}$$

8.2.2. Theorem

Let E be a countable bounded set then E is measurable and $m(E) = 0$ (23)

Proof: We consider a countable set E defined as

$$E = \{c_1, c_2, c_3, \ldots \ldots c_n, \ldots\}$$

where $a \leq c_i \leq b$ for all $i = 1, 2, 3, \ldots\ldots$

We enclose each c_i by an open interval $(c_i - h_i, c_i + h_i)$ such that each of these sub-intervals are pairwise disjoint. Then we have

$$m_0(E) \leq \sum_{i=1}^{\infty} l\{(c_i - h_i, c_i + h_i)\}$$

$$\leq 2\sum_{i=1}^{\infty} h_i$$

As each h_i is arbitrary, for any $\varepsilon > 0$ take

$$h_i = \frac{\varepsilon}{2^{i+1}}$$

Then

$$m_0(E) \leq \varepsilon \sum_{i=1}^{\infty} \frac{1}{2^i}$$

Since $\varepsilon > 0$ is arbitrary it follows that

$$m_0(E) = 0$$

Now the result follows from Theorem 8.2.1.

This theorem implies that any finite subsets of the real line R is measurable and its measure is zero. The set of rational numbers in any interval $[a, b]$ is measurable and its measure is zero.

8.2.3. Theorem

Let E be a set in $[a, b]$ and E^c be its complement, then

$$m_0(E) + m_c(E^c) = b - a \tag{24}$$

Proof: We consider an open set G_i in $[a, b]$ which contains E. Its complement G_i^c will be a closed set such that $G_i^c \subset E^c$.

This implies that

$$m_c(E^c) \geq l\{G_i^c\}$$

or

$$m_c(E^c) + l\{G_i\} \geq l\{G_i^c\} + l\{G_i\}$$

or

$$m_c(E^c) + l\{G_i\} \geq b - a \tag{25}$$

Taking infimum for all i we get

$$m_c(E^c) + m_0(E) \geq b - a$$

Similarly, we consider a closed set F_i contained in E^c.

We have

$$m_0(E) + l\{F_i\} \leq l\{F_i^c\} + l\{F_i\}$$

The value of right hand side is $b - a$ and taking supremum yields

$$m_0(E) + m_0(E^c) \leq b - a \tag{26}$$

We obtain the result from (25) and (26). On the above lines we can prove the similar result

$$m_0(E^c) + m_c(E) = b - a \tag{27}$$

Therefore if E is measurable, then

$$m(E^c) + m(E) = b - a \tag{28}$$

8.2.4. Theorem

If E_1 and E_2 are measurable subsets of $[a, b]$, then (i) $E_1 \cup E_2$ and $E_1 \cap E_2$ are measurable,

$$\text{(ii)} \quad m(E_1) + m(E_2) = m(E_1 \cup E_2) + m(E_1 \cap E_2) \tag{29}$$

Proof: First we prove that

$$m_0(E_1 \cup E_2) + m_0(E_1 \cap E_2) \leq m_0(E_1) + m_0(E_2) \tag{30}$$

We consider the open sets G_1 and G_2 containing E_1 and E_2 respectively, such that

$$m(G_1) - m_0(E_1) < \varepsilon/2, \; m(G_2) - m_0(E_2) < \varepsilon/2,$$

where ε is arbitrary. Then

$$\begin{aligned} m_0(E_1 \cup E_2) + m_0(E_1 \cap E_2) &\leq m(G_1 \cup G_2) + m(G_1 \cap G_2) \\ &= m(G_1) + m(G_2) \\ &\leq m_0(E_1) + m_0(E_2) + \varepsilon \end{aligned}$$

Since ε is arbitrary, this proves (30). As E_1 and E_2 are measurable we have

$$\begin{aligned} m(E_1) + m(E_2) &\geq m_0(E_1 \cup E_2) + m_0(E_1 \cap E_2) \\ &\geq m_c(E_1 \cup E_2) + m_c(E_1 \cap E_2) \end{aligned} \tag{31}$$

But on the above lines we can also prove that

$$m_c(E_1 \cup E_2) + m_c(E_1 \cap E_2) \geq m(E_1) + m(E_2) \tag{32}$$

Inequalities (30) and (31) imply that

$$\begin{aligned} m(E_1) + m(E_2) &\geq m_0(E_1 \cup E_2) + m_0(E_1 \cap E_2) \geq \\ & m_c(E_1 \cup E_2) + m_c(E_1 \cap E_2) \geq m(E_1) + m(E_2) \end{aligned} \tag{33}$$

This means that

$$m_0(E_1 \cup E_2) + m_0(E_1 \cap E_2) = m_c(E_1 \cup E_2) + m_c(E_1 \cap E_2) \tag{34}$$

We get

$$m_0(E_1 \cup E_2) = m_c(E_1 \cup E_2)$$

and

$$m_0(E_1 \cap E_2) = m_c(E_1 \cap E_2)$$

This means that $E_1 \cup E_2$ and $E_1 \cap E_2$ are measurable and this also proves the theorem.

Also if E_1 and E_2 are disjoint then

$$m(E_1 \cup E_2) = m(E_1) + m(E_2) \tag{35}$$

This result can be generalised to a countable collection of sets $\{E_i\}$, $i = 1, 2, 3, \ldots$, which are pairwise disjoint. This gives

$$m\left[\bigcup_{i=1}^{\infty} E_i\right] = \sum_{i=1}^{\infty} m(E_i) \tag{36}$$

Now we prove theorems on the intersection of measurable sets.

8.2.5. Theorem

Let $\{E_i\}$ be a countable collection of sets such that $E_{i+1} \subset E_i$ for each $i = 1, 2, 3, \ldots$

Let $m(E_i)$ bę finite for at least one i. Then

$$m\left(\bigcap_{i=1}^{\infty} E_i\right) = \lim_{n\to\infty} m(E_n)$$

Proof: Let r be the least value of i such that $m(E_r)$ is finite. Then $m(E_i) < \infty$ for all $i \geq r$. We put $E = \bigcap_{i=1}^{\infty} E_i$ and $F_i = E_i - E_{i+1}$. Then the sets F_i are measurable and pairwise disjoint, and $E_r - E = \bigcup_{i=r}^{\infty} F_i$

Therefore,

$$m(E_r - E) = \sum_{i=r}^{\infty} m(F_i)$$

$$= \sum_{i=r}^{\infty} m(E_i - E_{i+1})$$

We have

$$m(E_r) = m(E) + m(E_r - E)$$

and

$$m(E_i) = m(E_{i+1}) + m(E_i - E_{i+1}) \text{ for } i \geq r$$

Hence

$$m(E_r) - m(E) = \sum_{i=r}^{\infty} [m(E_i) - m(E_{i+1})]$$

$$= \lim_{n\to\infty} \sum_{i=r}^{n} [m(E_i) - m(E_{i+1})]$$

$$= \lim_{n\to\infty} \sum_{i=r}^{\infty} [m(E_i - E_{i+1})]$$

$$= \lim_{n\to\infty} [m(E_r - E_n)]$$

$$= m(E_r) - \lim_{n\to\infty} (E_n)$$

This gives
$$m(E) = \lim_{r\to\infty} m(E_n) \tag{37}$$

Hence the theorem.

We have the following theorem whose proof is left to the reader:

8.2.6. Theorem

If $E = \bigcap_{i=1}^{\infty} E_i$, where each E_i is bounded and measurable, then E is measurable.

8.2.7. Non-measurable Sets

There are sets in which the outer measure is different from the inner measure. Following is an example of a non-measurable set: We take a segment (interval) of real line of length π and bend it to form a circle of radius 1/2. Then we divide the points of the circumference into sets in such a way that, if P is a point of any particular set, then the points distant 1, 2, 3, ... from it in either direction along the circumference belong to the same set. Then we take one representative of each of those sets, forming a set E. Let E_n, E_{-n} be the sets obtained from E by rotation of n in the two directions round the circle. Then it is clear that the sets E, E_1, E_{-1}, E_2, E_{-2}, ... are disjoint and all together they fill the circumference. Moreover, they are congruent with one another, and if measurable must all have same measure. This measure cannot be zero as then the measure of the sum will be zero, which is not true. The sum of all the measures should be π. This measure cannot be positive, because the circumference in this case will have infinite measure and this is not possible. Thus E is not measurable.

8.2.8. Examples

(a) A singleton set is measurable and its measure is zero.

Let $\{x\}$ be the given set. Since

$$x \,\varepsilon\, I_n = \left(x - \frac{1}{2^{n+1}},\ x + \frac{1}{2^{n+1}} \right)$$

and $l(I_n) = \frac{1}{2^n}$.

$$m_0(\{x\}) = \lim \frac{1}{2^n} = 0.$$

Thus the result holds.

(b) Prove that if E is the set given by

$$E = \{0, 1, 1/2, 1/3, \ldots\}$$

Then $m(E) = 0$

Solution: We consider the closed interval $I_{0,1}$. Then

$$m(I_{0,1}) = l(I_{0,1}) = 1$$

$$I_{0,1} - E = \bigcup_{n=1}^{\infty} I_n$$

Where I_n is an open interval given by

$$I_n = (1/n+1,\ 1/n).\ l(I_n) = 1/n - 1/n + 1 = 1/n(n+1)$$

Therefore $m(I_{0,1} - E) = \sum_{n=1}^{\infty} \frac{1}{n(n+1)} = 1$

But $m(I_{0,l} - E) = m(I_{0,1}) - m(E)$

$= 1 - m(E)$

Hence $m(E) = 0$.

(c) There exist uncountable sets of measure zero. In fact Cantor's ternary set discussed in Chapter 2(2.3.9) is uncountable and its measure is zero. Recall that Cantor's ternary set is the set P defined by

$$P = \left(I - \bigcup_{n=1}^{\infty} I_n \right) = \left(\bigcup_{n=1}^{\infty} I_n \right)^c$$

where

$$I = [0,\ 1],\ I_1 = \left(\frac{1}{3}, \frac{2}{3}\right),$$

$$I_2 = \left(\frac{1}{3^2}, \frac{2}{3^2}\right) \cup \left(\frac{7}{3^2}, \frac{8}{3^2}\right)$$

$$\cdots \quad \cdots \quad \cdots \quad \cdots$$

Since P is a closed set, it is measurable. We have

$$m\left(\bigcup_{n=1}^{\infty} I_n\right) = \frac{1}{3} + \frac{2}{3^2} + \frac{2^2}{3^3} + \ldots$$

$$= \frac{\frac{1}{3}}{1 - \frac{2}{3}} = 1.$$

So $m(P) = m([0,\ 1]) - m\left(\bigcup_{n=1}^{\infty} I_n\right)$

$= 1 - 1 = 0.$

We now show that the set P is uncountable. Assume, to the contrary, that P is countable. To be precise, let

$$P = \{x_1, x_2, \ldots\}.$$

We now represent the points of P by their ternary expansions. Note that the ternary expansion of a number is the expansion in the scale of three of the form $\sum_{k=1}^{\infty} \frac{a^k}{2^k}$ where each $a^k = 0$ or 2. Let the ternary expansion of x_i be given by

$$x_i = 0.\ a_{i1}\ a_{i2}\ a_{i3} \ldots,\ i = 1, 2, 3, \ldots$$

where each $a_{ij} = 0$ or 2. Now construct a point y as

$$y = 0.\ b_1\ b_2\ b_3 \ldots$$

where for $i = 1, 2, 3, \ldots$

$$b_i = \begin{cases} 0, & a_{ii} = 2, \\ 2, & a_{ii} = 0. \end{cases}$$

Since $b_i = 0$ or 2, $b_i \varepsilon P$. But $y \neq x_i$ for any i. This contradicts our assumption that P is countable and completes the proof.

8.3 MEASURABLE FUNCTIONS

Let f be an extended real-valued function with domain $U \subset [a, b]$. The function f is said to be measurable, if for every real number r any one of the sets E_i ($i = 1, 2, 3, 4$) defined as

$$E_1 = \{x : f(x) < r\} \tag{38a}$$

$$E_2 = \{x : f(x) \leq r\} \tag{38b}$$

$$E_3 = \{x : f(x) > r\} \tag{38c}$$

$$E_4 = \{x : f(x) \geq r\} \tag{38d}$$

is measurable.

It can be proved that any one of the above implies the other three.

We have $E_1 \cup E_4 = U$. Thus E_4 is the complement of E_1, hence measurable for E_1 is measurable. The same is true for the sets E_2 and E_3. Now the set E_4 can also be expressed as

$$E_4 = \bigcap_{i=1}^{\infty} H_i \tag{39}$$

Where $H_i \left\{x : f(x) > r - \frac{1}{i}\right\}$

Thus each H_i is of the type E_3.

Therefore, if H_i is measurable, then E_4 is also measurable. Hence the measurability of E_3 implies the measurability of E_4. In this manner we can show that any one of the definitions in (38) implies the other three.

Definition: Two functions f and g are said to be equal almost everywhere, written as $f = g$ a.e. if the set $E = \{x : f(x) \neq g(x)\}$ has zero measure.

Some well known properties of measurable functions are listed below in the form of a theorem.

8.3.1. Theorem

(i) If f is measurable and c is a constant, then the functions $f + c$ and cf are measurable.

(ii) If f is measurable, then the set $\{x : f(x) = \alpha\}$ is measurable for each real number α or for $\alpha = \infty$.

(iii) If f is measurable, then $|f|$ is measurable.

(iv) If f and g are measurable, then $f + g$ and $f - g$ are measurable.

(v) If f and g are measurable then the set

$$E = \{x : f(x) > g(x)\}$$

is measurable.

(vi) If f and g are measurable functions, then fg is measurable.

(vii) If f and g are measurable and $g \neq 0$, then f/g is measurable.

(viii) If f and g are measurable, then the functions max (f, g) and min (f, g) are measurable.

(ix) If $\{f_n\}$ is a sequence of measurable functions, each defined on $[a, b]$ such that the sequence $\{f_n(x)\}$ is bounded for every $x \varepsilon [a, b]$, then the functions

$$G(x) = \sup f_n(x),$$

$$g(x) = \inf f_n(x),$$

$$H(x) = \overline{\lim_{n\to\infty}} f_n(x), \text{ and}$$

$$h(x) = \underline{\lim_{n\to\infty}} f_n(x),$$

are all measurable.

(x) The limit of a convergent sequence of measurable functions is measurable.

(xi) Every continuous function is measurable.

(xii) If $f = g$ a, e, and f is measurable, then g is measurable.

8.3.2. Examples

(a) Prove that all constant functions are measurable.

Solution: Let f be a real-valued function such that

$$f(x) = K \text{ for all real } x$$

We consider a real number r and a set S such that

$$S = \{x : f(x) \geq r\}$$

If $K \geq r$, then $S = R$, which is measurable.

Similarly if $K < r$, then $S = \varnothing$ which is again measurable. Hence f is measurable for all K.

(b) Let h be defined by

$$h(x) = g(f(x))$$

where f is measurable and g is monotonically increasing. Then prove that h is measurable.

Solution: We have that the set

$$S = \{x : f(x) \geq c\},\ c \text{ real, is measurable. Now}$$

$f(x) \geq c \Rightarrow g(f(x)) \geq K$ where $K = g(c)$ as g is monotonic increasing. Therefore, S can also be written as

$$S = \{x : h(x) \geq K\}$$

But S is measurable. Hence h is a measurable function.

8.4 LEBESGUE INTEGRAL

Having discussed measurable sets and measurable functions, we can now define the *Lebesgue integral* of a bounded function f with its domain on the real line.

We consider an interval $[a, b]$. Divide $[a, b]$ into measurable subsets $E_1, E_2, \ldots E_n$, such that

$$m(E_i \cap E_j) = 0, \quad i \neq j$$

and

$$\bigcup_{i=1}^{n} E_i = [a, b] \tag{39}$$

We denote this division of $[a, b]$ by D and call it a measurable partition. We can have a refinement D^* of $[a, b]$ where

$$D^* = \{E_1^*, E_2^* \ldots, E_n^*\}$$

such that for every E_i of D there is a measurable set E_j^* of D^* satisfying $E_j^* \subset E_i$. However, we shall denote a general partition by D where $D = \{E_1, E_2, \ldots E_n\}$. (40)

We define

$$l_i = \inf_{x \in E_i} f(x) \tag{41}$$

and

$$L_i = \sup_{x \in E_i} f(x) \tag{42}$$

for $i = 1, 2, \ldots, n$.

Then we define the lower Lebesgue sum $s(f, D)$ and upper Lebesgue sum $S(f, D)$ as follows:

$$s(f, D) = \sum_{i=1}^{n} l_i \, m(E_i) \qquad (43)$$

$$S(f, D) = \sum_{i=1}^{n} L_i \, m(E_i) \qquad (44)$$

It is easy to show that $s(f, D) \leq S(f, D)$. (45)

Further, if D_2 is any refinement of D_1, then it is easy to show that

$$s(f, D_1) \leq s(f, D_2) \qquad (46)$$

$$S(f, D_1) \geq S(f, D_2) \qquad (47)$$

Let $\mathcal{D}$ denote the set of all measurable partitions of $[a, b]$. We now define

$$\underline{\int_a^b} f dx = \sup_{D \in \mathcal{D}} s(f, D) \qquad (48)$$

and

$$\overline{\int_a^b} f dx = \inf_{D \in \mathcal{D}} S(f, D) \qquad (49)$$

$\underline{\int_a^b} f dx$ and $\overline{\int_a^b} f dx$ are respectively called the lower Lebesgue integral and upper Lebesgue integral of f in $[a, b]$.

A bounded function f is said to be Lebesgue integrable over $[a, b]$ if the lower Lebesgue integral of f is equal to its upper Lebesgue integral, i.e. if

$$\underline{\int_a^b} f dx = \overline{\int_a^b} f dx. \qquad (50)$$

The common value is called the Lebesgue integral of f and is denoted by

$$\int_a^b f dx$$

We write $L_1[a, b]$ or simply $L[a, b]$ to denote the set of all Lebesgue integrable functions over $[a, b]$.

8.4.1. Theorem

Let f be a bounded function which is Riemann integrable in the interval $[a, b]$. Then f is Lebesgue integrable on $[a, b]$ and

$$R\int_a^b f dx = L\int_a^b f dx \tag{51}$$

Proof: Let P be, a partition of $[a, b]$ into subintervals with end points x_0 $(=a)$, x_1, x_2, .. x_{n-1}, x_n $(= b)$, i.e.

$$P = \{x_0, x_1, x_2, \ldots, x_n\}$$

We also define a division D of $[a, b]$, where

$$D = \{E_1, E_2, E_3, \ldots, E_{2n}\}$$

such that

$$E_1 \cup E_2 = [x_0, x_1], E_3 \cup E_4 = [x_1, x_2], \ldots$$
$$\ldots E_{2n-1} \cup E_{2n} = [x_{n-1}, x_n]$$

where all E_i's $(i = 1, 2, \ldots, 2n)$ are pairwise disjoint.

We can also write P as a division D with its members as the intervals $I_1, I_2, \ldots, I_n$ where

$$I_i = [x_{i-1}, x_i] \text{ for all } i = 1, 2, \ldots, n.$$

Now if $L(f, P)$ and $U(f, P)$ denote the lower and upper Riemann sums respectively, then, using (43), (44) and (45), we get

$$L(f, P) \le s(f, D) \le S(f, D) \le U(f, P) \tag{52}$$

$$R\int_{\underline{a}}^{b} f dx \le L\int_{\underline{a}}^{b} f dx \le L\int_a^{\bar{b}} f dx \le R\int_a^{\bar{b}} f dx \tag{53}$$

If $f \in R[a, b]$, then

$$R\int_{\underline{a}}^{b} f dx = R\int_a^{\bar{b}} f dx \tag{54}$$

Hence

$$R\int_{\underline{a}}^{b} f dx = L\int_{\underline{a}}^{b} f dx = L\int_a^{\bar{b}} f dx = R\int_a^{\bar{b}} f dx \tag{55}$$

or $\quad f \in L[a, b]$, and

$$R\int_a^b f dx = L\int_a^b f dx.$$

This proves the theorem.

We now prove that every bounded measurable function is Lebesgue integrable. In order to prove this result we use the following theorem whose proof is beyond the scope of this book.

8.4.2. Theorem

A bounded function f is Lebesgue integrable over $[a, b]$ if and only if for every $\varepsilon > 0$ there exists a measurable partition D of $[a, b]$ such that

$$S(f, D) - s(f, D) < \varepsilon.$$

8.4.3. Theorem

Every bounded measurable function on $[a, b]$ is Lebesgue integrable on $[a, b]$.

Proof: Let f be bounded. Then there exist m and M such that $f(x) \varepsilon [m, M]$ for all $x \varepsilon [a, b]$. Let $\varepsilon > 0$ be given. Then there exist a finite number of points $y_0, y_1, \ldots, y_n$ such that

$$m = y_0 < y_1 < y_2 < \ldots < y_n = M$$

and

$$y_k - y_{k-1} < \frac{\varepsilon}{b - a}$$

for $k = 1, 2, \ldots, n$. Let

$$E_k = \{x \varepsilon [a, b] : y_{k-1} \leq f(x) < y_k\}.$$

Since f is measurable, each E_k is measurable. Thus

$$D = \{E_1, E_2, \ldots, E_n\}$$

is a measurable partition of $[a, b]$. Now

$$\sup_{x \varepsilon E_k} f(x) \leq y_k, \inf_{x \varepsilon E_k} f(x) \geq y_{k-1}.$$

Therefore

$$S(f, D) = \sum_{k=1}^{n} \left(\sup_{x \varepsilon E_k} f(x) \right) m(E_k)$$

$$\leq \sum_{k=1}^{n} y_k \, m(E_k).$$

$$s(f, D) = \sum_{k=1}^{n} \left(\inf_{x \in E_k} f(x) \right) m(E_k)$$

$$\geq \sum_{k=1}^{n} y_{k-1}\, m(E_k).$$

Now

$$S(f, D) - s(f, D) \leq \sum_{k=1}^{n} (y_k - y_{k-1})\, m(E_k)$$

$$< \frac{\varepsilon}{(b-a)} \sum_{k=1}^{m} m(E_k)$$

$$< \varepsilon$$

Hence f is Lebesgue integrable and this completes the proof.

8.4.5. Simple Properties of Lebesgue Integral of Bounded Functions

Theorem

Let f and g be bounded functions and $f \in L[a, b]$, $g \in L[a, b]$. Then following properties are satisfied:

(i) $f \leq g \Rightarrow \int_a^b f dx \leq \int_a^b g dx$,

(ii) $\int_a^b cf\, dx = c \int_a^b f dx$, c is a constant,

(iii) $\int_a^b (f + g)\, dx = \int_a^b f dx + \int_a^b g dx$,

(iv) $\int_a^b f dx = \int_a^c f dx + \int_c^b f dx$ for $a \leq c \leq b$.

(v) If f is bounded and $f \in L[a, b]$, $f = g$ a.e. on $[a, b]$ and g is bounded on $[a, b]$, then $g \in L[a, b]$ and

$$\int_a^b g dx = \int_a^b f dx$$

(vi) If f and g are bounded, $f \varepsilon L[a, b]$ and $g \varepsilon L[a, b]$, then

$$f(x) \leq g(x) \text{ a.e.} \Rightarrow \int_a^b f dx \leq \int_a^b g dx.$$

(vii) If a bounded function f is Lebesgue integrable on $[a, b]$, then $|f|$ is Lebesgue integrable and

$$\left| \int_a^b f dx \right| \leq \int_a^b |f| \, dx.$$

8.4.6. Example

Evaluate $\int_a^b f(x) dx$

where f is defined by

$$f(x) = \begin{cases} K_1, & x \varepsilon I_{a, b}, \ x \text{ rational}, \\ K_2, & x \varepsilon I_{a, b}, \ x \text{ irrational}, \ (K_1 < K_2) \end{cases}$$

Solution: We have $f(x) \varepsilon [K_1, K_2]$. We choose a partition D of $[a, b]$ where $D = \{E_1, E_2, ... E_n\}$.
Let

$y_0, y_1, y_2, ..., y_n$ be real numbers such that

$$K_1 - \epsilon = y_0 < y_1 < y_2 < ... < y_{n-1} < y_n = K_2 + \epsilon, \ \epsilon > 0$$

Here $E_1, E_2, ... E_n$ are subsets of $[a, b]$ such that

$$E_i = \{x : y_{i-1} \leq f(x) < y_i\} \text{ for } i = 1, 2, ..., n.$$

Then E_1 is the set of all rational numbers in $[a, b]$ while E_n is the set of all irrational numbers in $[a, b]$. Hence $m(E_1) = 0$, $m(E_n) = (b - a) - m(E_1) = (b - a)$ and $m(E_i) = 0$ for $i = 1, 2, ..., n - 1$.
Therefore

$$S(f, D) = \sum_{i=1}^{n} \sup_{x \varepsilon E_i} f(x) \, m(E_i)$$

$$\leq y_n \, m(E_n) = (b - a)(K_2 + \varepsilon)$$

$$s(f, D) = \sum_{i=1}^{n} \inf_{x \varepsilon E_i} f(x)\, m(E_i) \geq y_n\, m(E_n) =$$

$$= (b - a)(K_2 - \varepsilon)$$

Thus for any partition D we have $\inf_D S(D, f) = \sup_D s(D, f)$

$$= K_2 (b - a).$$

Hence

$$\int_a^b f dx = K_2 (b - a).$$

We can even take a partition D of $[a, b]$ such that $D = \{E_1, E_2\}$ where

$$E_1 = \{ x \varepsilon [a, b] : x \text{ is rational } \},$$
$$E_2 = \{ x \varepsilon [a, b] : x \text{ is irrational}\}.$$

Then

$$s(f, D) = \inf_{x \varepsilon E_1} f(x)\, m(E_1) + \inf_{x \varepsilon E_2} f(x)\, m(E_2)$$

$$= K_1\, m(E_1) + K_2\, m(E_2)$$

$$= 0.\, K_1 + K_2 (b - a) = K_2 (b - a)$$

$$S(f, D) = \sup_{x \varepsilon E_1} f(x)\, m(E_1) + \sup_{x \varepsilon E_2} f(x)\, m(E_2)$$

$$= K_1 .\, 0 + K (b - a)$$

$$= K_2 (b - a).$$

Therefore

$$\sup_D s(f, D) = \inf_D S(f, D) = K_2 (b - a),$$

hence f is Lebesgue integrable and

$$\int_a^b f dx = K_2 (b - a).$$

8.4.7. Lebesgue Integral for Unbounded Functions

Let f be a non-negative measurable function defined on $[a, b]$. We define a function $F : [a, b] \times N \rightarrow R$ by

$$F(x, n) = \begin{cases} f(x), & 0 \le f(x), \le n, \\ n \;, & f(x) > n. \end{cases}$$

Now $F(x, n) = \min(f(x), n)$. Thus F is a bounded measurable function for each $n \in N$ and hence F is Lebesgue integrable.
Let

$$\int_a^b F(x, n)\, dx$$

denote the Lebesgue integral of F. If

$$\lim_{n \to \infty} \int_a^b F(x, n)\, dx$$

exists and is finite, then we say that f is Lebesgue integrable and

$$\int_a^b f dx = \lim_{n \to \infty} \int_a^b F(x, n)\, dx.$$

If the limit does not exist finitely, then we say that f is not Lebesgue integrable.

We now extend the definition of Lebesgue integral to any real-valued measurable function defined on any measurable subset of the real line R.

8.4.8. Definition

Let f be a real measurable function defined on a measurable subset A of R. We define two functions f^+ and f^- as follows:

$$f^+(x) = \max(f(x), 0),$$
$$f^-(x) = \max(-f(x), 0) = -\min(f(x), 0)$$

Observe that

$$f = f^+ - f^-, \; |f| = f^+ + f^- ,$$

f^+ and f^- are non-negative functions.

Definition: Let f be a measurable function defined on a measurable set A and let $\int_A f^+\, dx < \infty$, $\int_A f^-\, dx < \infty$. Then f is Lebesgue integrable and

$$\int_A f dx = \int_A f^+ dx - \int_A f^- dx.$$

If

$$\int_A f^+ dx = \infty \text{ and } \int_A f^- dx = \infty,$$

then $\int_A f dx$ is undefined.

Observe that $\int_A f dx$ is finite iff $\int_A f^+ dx$ and $\int_A f^- dx$ are finite and this is true iff

$$\int_A |f| dx = \int_A f^+ dx + \int_A f^- dx \text{ is finite.}$$

8.4.9. Theorem

Let f be a measurable function defined on $[a, b]$. Then $f \varepsilon L[a, b]$ iff $|f| \varepsilon L[a, b]$ and then

$$\left| \int_a^b f dx \right| \leq \int_a^b |f| dx$$

Proof: Let f be measurable and $|f| \varepsilon L[a, b]$. We have

$$0 \leq f^+(x) \leq |f(x)|.$$

$$0 \leq f^-(x) \leq |f(x)|.$$

This implies that f^+ and f^- are Lebesgue integrable and hence f is Lebesgue integrable.

Conversely, let $f \varepsilon L[a, b]$. Then f^+ and f^- are Lebesgue integrable. But

$$|f| = f^+ + f^-$$

Hence $|f| \varepsilon L[a, b]$. Now

$$f \leq |f|, -f \leq |f|$$

Therefore

$$\int_a^b f dx \leq \int_a^b |f| dx, \quad -\int_a^b f dx \leq \int_a^b |f| dx$$

Thus

$$\left| \int_a^b f dx \right| \leq \int_a^b |f| dx$$

and this completes the proof.

8.4.10. Theorem

Let $f \varepsilon L_1 [a, b]$. Then given $\varepsilon > 0$ there exists a $\delta > 0$ such that

$$\left| \int_E f dx \right| < \varepsilon$$

where $E \subset [a, b]$ is measurable with $m(E) < \delta$.

Proof: We first assume that f is a non-negative function. For any positive integer n, define

$$F(x, n) = \begin{cases} f(x) & \text{if } 0 \leq f(x) < n, \\ 0 & \text{if } f(x) \geq n. \end{cases}$$

Then, given $\varepsilon > 0$, there exists a positive integer N such that

$$\int_a^b f(x)\, dx - \int_a^b F(x, N)\, dx < \frac{\varepsilon}{2}$$

i.e.

$$\int_a^b \{f(x) - F(x, N)\}\, dx < \frac{\varepsilon}{2}. \tag{56}$$

Let $\delta > 0$ be such that $\delta < \dfrac{\varepsilon}{2N}$.

Let $E \subset [a, b]$ be measurable and $m(E) < \delta$. Then we have

$$\int_E F(x, N)\, dx \leq \int_E N dx = N\, m(E)$$

$$< N\delta < n \frac{\varepsilon}{2N} = \frac{\varepsilon}{2}. \tag{57}$$

From (56) and (57) we get

$$\int_E f(x)dx = \int_E \{f(x) - F(x, N)\}dx + \int_E F(x, N)dx$$

$$< \frac{\varepsilon}{2} + \frac{\varepsilon}{2} = \varepsilon.$$

Now, if f is an arbitrary Lebesgue integrable function on $[a, b]$, then

$$f = f^+ - f^-$$

where f^+ and f^- are non-negative and Lebesgue integrable. Hence for a given $\varepsilon > 0$, there exists a $\delta_1 > 0$ such that

$$\int_E f^+ \, dx < \frac{\varepsilon}{2} \text{ where } m(E) < \delta_1.$$

Similarly there exists $\delta_2 > 0$ such that

$$\int_E f^- \, dx < \frac{\varepsilon}{2} \text{ for } m(E) < \delta_2.$$

Let $\delta = \min(\delta_1, \delta_2)$. Then we get

$$\left| \int_E f dx \right| \le \int_E |f| dx$$

$$= \int_E f^+ \, dx + \int_E f^- \, dx$$

$$< \frac{\varepsilon}{2} + \frac{\varepsilon}{2} = \varepsilon,$$

and this completes the proof.

8.4.11. Lebesgue Bounded Convergence Theorem

Let $\{f_n\}$ be a sequence of measurable functions defined on a measurable set $E \subset [a, b]$. Suppose

(i) there exists a constant M with $|f_n(x)| \le M$ for all n and x; and

(ii) $\lim_{n\to\infty} f_n(x) = f(x)$.

Then

$$\lim_{n\to\infty} \int_E f_n(x) \, dx = \int_E f(x) \, dx.$$

Proof: Given that

$$\lim_{n\to\infty} f_n(x) = f(x) \text{ and } |f_n(x)| \le M.$$

Therefore $|f(x)| \le M$. Further f, being the limit of a sequence of measurable functions, is measurable. Hence f is Lebesgue integrable. It is now enough to show that

$$\lim_{n\to\infty} \int_E |f_n(x) - f(x)|\, dx = 0.$$

For given $\varepsilon > 0$, define a partition of E into disjoint measurable sets as follows:

$$E_k = \{x : |f_{k-1} - f| \ge \varepsilon,\ |f_n - f| < \varepsilon,\ n \ge k\},\ k = 1, 2, \ldots$$

Thus

$$E_1 = \{x : |f_n - f| < \varepsilon,\ n = 1, 2, \ldots\},$$

$$E_2 = \{x : |f_1 - f| \ge \varepsilon,\ |f_n - f| < \varepsilon,\ n = 2, 3, \ldots\}$$

Clearly

$$E = \bigcup_{k=1}^{\infty} E_k = \left(\bigcup_{k=1}^{n} E_k\right) \cup \left(\bigcup_{k=n+1}^{\infty} E_k\right)$$

$$= P_n \cup Q_n \text{ (say)}$$

Hence

$$m(E) = m(P_n \cup Q_n) = m(P_n) + m(Q_n).$$

Now

$$\int_E |f_n - f|\, dx = \int_{P_n} |f_n - f|\, dx + \int_{Q_n} |f_n - f|\, dx$$

For each n,

$$|f_n - f| < \varepsilon \text{ on } P_n \text{ and}$$

$$|f_n - f| \le |f_n| + |f| \le M + M = 2M \text{ on } Q_n.$$

Thus

$$\int_E |f_n - f|\, dx < \varepsilon\, m(P_n) + 2M\, m(Q_n)$$

Letting $n \to \infty$ we have

$$m(P_n) \to m(A),\ m(Q_n) \to 0.$$

Further, since $\varepsilon > 0$ is arbitrary we obtain the desired result.

8.4.12. Lebesgue Dominated Convergence Theorem

Let $\{f_n\}$ be a sequence of measurable functions on $[a, b]$. Suppose

(i) there exists a function $g \varepsilon L_1 [a, b]$ such that

$$| f_n(x) | \leq g(x) \ a.e.$$

(ii) $\lim\limits_{n \to \infty} f_n(x) = f(x) a.e.$

Then $f \varepsilon L_1 [a, b]$ and

$$\lim_{n \to \infty} \int_a^b f_n \, dx = \int_a^b f dx.$$

Proof: Since f is the limit of a sequence of measurable functions, it is measurable. Again

$$| f_n(x) | \leq g(x) \text{ and } \lim_{n \to \infty} f_n(x) = f(x) \text{ a.e. on } [a, b].$$

Hence $\quad | f(x) | \leq g(x)$ a.e. on $[a, b]$.

Now, since g is Lebesgue integrable, $| f |$ is also so and, hence, so is f.

For given $\varepsilon > 0$ and for any positive integer N define

$$E_N = \{ x : | f_n(x) - f(x) | < \frac{\varepsilon}{2(b-a)}, \quad n \geq N, x \varepsilon [a, b]\},$$

Now $E_N \subset E_{N+1}$ for all N. Since $\lim\limits_{n \to \infty} f_n(x) = f(x)$ a.e. on $[a, b]$, it follows that

$$[a, b] = \left(\bigcup_{n=1}^{\infty} E_N \right) \cup E \text{ where } m(E) = 0.$$

Therefore $m\left(\bigcup\limits_{n+1}^{\infty} E_N \right) = b - a$

Also $m\left(\bigcup\limits_{N=1}^{\infty} E_N \right) = \lim\limits_{N \to \infty} m\left(\bigcup\limits_{n=1}^{N} E_n \right)$

$$= \lim_{n \to \infty} m(E_N)$$

[since $E_1 \subset E_2 \ldots. \subset E_{N-1} \subset E_N \ldots$].

Therefore

$$b - a = \lim_{N \to \infty} m(E_N). \tag{58}$$

Given that g is Lebesgue integrable. Therefore, for every $\varepsilon > 0$ there exists a $\delta > 0$ such that

$$\int_E g\,dx < \frac{\varepsilon}{4} \text{ for } m(E) < \delta.$$

We have

$$\int_E |f_n|\,dx \le \int_E g\,dx < \frac{\varepsilon}{4} \text{ for } m(E) < \delta \tag{59}$$

and hence

$$\int_E |f|\,dx \le \int_E g\,dx < \frac{\varepsilon}{4} \text{ for } m(E) < \delta \tag{60}$$

From (58) it follows that there exists a positive integer M such that

$$b - a - m(E_N) < \delta \text{ for } N \ge M. \tag{61}$$

or
$$m(E_N^c) < \delta \text{ for } N \ge M.$$

Now for $n \ge M$,

$$\int_a^b |f_n - f|\,dx = \int_{E_M} |f_n - f| + \int_{E_M^c} |f_n - f|$$

$$< \frac{\varepsilon\, m(E_M)}{2(b-a)} + \int_{E_M^c} |f_n| + \int_{E_M^c} f$$

$$< \frac{\varepsilon\, m(E_M)}{2(b-a)} + \frac{\varepsilon}{4} + \frac{\varepsilon}{4}$$

or $\quad \left|\int_a^b f_n - f\right| \le \int_a^b |f_n - f| < \varepsilon \text{ for } n \ge M$

or, $\quad \left|\int_a^b f_n - \int_a^b f\right| < \varepsilon \text{ for } n \ge M$

or, $\quad \lim_{n\to\infty} \int_a^b f_n = \int_a^b f$

and this completes the proof.

8.4.13. Theorem (Fatou's Lemma)

Let $\{f_n\}$ be sequence of non-negative measurable functions on $[a, b]$ such that

$\lim_{n\to\infty} f_n(x) = f(x)$ for all $x \in [a, b]$. Then

$$\int_a^b f\,dx \le \lim_{n\to\infty} \int_a^b f_n\,dx.$$

Proof: For any positive integer m define a function F_n on $[a, b]$ by

$$F_n(x, m) = \begin{cases} f_n(x), & 0 \le f_n(x) \le m\,; \\ m, & f_n(x) > m \end{cases}$$

Therefore $F_n(x, m) = \min(f_n(x), m)$ and thus each $F_n(x, m)$ is bounded by m. It is also measurable and hence is Lebesgue integrable on $[a, b]$. Also

$$\begin{aligned} \lim_{n\to\infty} F_n(x, m) &= \lim_{n\to\infty} \min(f_n(x), m) \\ &= \min\left(\lim_{n\to\infty} f_n(x), m\right) \\ &= \min(f(x), m) \\ &= F(x, m) \end{aligned}$$

Therefore, by the Lebesgue dominated convergence theorem, $F(x, m)$ is Lebesgue integrable and

$$\lim_{n\to\infty} \int_a^b F_n(x, m)\,dx = \int_a^b F(x, m)\,dx.$$

Now $F_n(x, m) \le f_n(x)$. Therefore

$$\begin{aligned} \int_a^b F(x, m)\,dx &= \lim_{n\to\infty} \int_a^b F_n(x, m)\,dx \\ &= \underline{\lim}_{n\to\infty} \int_a^b F_n(x, m)\,dx \\ &\ge \underline{\lim}_{n\to\infty} \int_a^b f_n\,dx \end{aligned}$$

But $F(x, m) \to f(x)$ as $m \to \infty$. Therefore

$$\int_a^b f(x)\, dx \leq \lim_{n\to\infty} f_n(x)\, dx$$

and this completes the proof.

8.5 THE SPACE L_2

We shall now introduce the Lebesgue space L_2. This space is of great importance and is frequently used in solving various problems. We shall use this concept in the next section (8.6) while discussing Fourier series.

Let f be a real-valued bounded measurable function defined on the closed interval $[a, b]$. Then f is said to be in $L_2[a, b]$ if

$$\int_a^b |f(x)|^2 dx$$

exists. In other words, $f \varepsilon L_2[a, b]$ iff $f^2 \varepsilon L_1[a, b]$. Thus $L_2[a, b]$ is the set of all square integrable functions. We shall write L_2 for $L_2[a, b]$ when there is no chance of confusion. We have

8.5.1 Theorem (Schwarz's Inequality)

If f and g belong to $L_2[a, b]$, then $fg \varepsilon L_1[a, b]$ and

$$\left|\int_a^b f(x)\, g(x)\, dx\right| \leq \left[\int_a^b |f(x)|^2 dx \int_a^b |g(x)|^2 dx\right]^{1/2}.$$

Proof: Since

$$0 \leq (f-g)^2 = f^2 + g^2 - 2|fg|,$$

it follows that

$$2|fg| \leq f^2 + g^2.$$

Let $f, g \varepsilon L_2$. Then

$$2\left|\int_a^b f(x)\, g(x)\, dx\right| \leq 2\int_a^b |f(x)\, g(x)|\, dx$$

$$\leq \int_a^b |f(x)|^2\, dx \int_a^b |g(x)|^2 dx.$$

Hence $fg \in L_1$. Therefore the integral

$$\int_a^b \{\lambda f(x) + \mu g(x)\}^2 dx$$

$$= \lambda^2 \int_a^b \{f(x)\}^2 dx + 2\lambda\mu \int_a^b f(x) g(x) dx + \mu^2 \int_a^b \{g(x)\}^2 dx$$

exists for all values of λ and μ, and is never negative. But the necessary and sufficient condition that

$$a\lambda^2 + 2h\lambda\mu + b\mu^2$$

is not negative is that

$$h^2 \le ab,\ a \ge 0,\ b \ge 0.$$

Therefore it follows that

$$\left[\int_a^b f(x) g(x)\, dx\right]^2 \le \int_a^b |f(x)|^2 dx \int_a^b |g(x)|^2 dx$$

and this completes the proof.

8.5.2. Theorem (Minkowski's Inequality)

Let $f, g \in L_2$. Then $f + g \in L_2$ and

$$[\int |f(x) + g(x)|^2 dx]^{1/2} \le [\int |f(x)|^2 dx^{1/2}] + [\int |g(x)|^2 dx]^{1/2}.$$

Proof: $\int |f+g|^2 dx$

$$\le \int |f|\,|f+g|\,dx + \int |g|\,|f+g|\,dx$$

$$\le \int |f|^2 dx]^{1/2} [\int |f+g|^2 dx]^{1/2}$$

$$+ [\int |g|^2 dx]^{1/2} [\int |f+g|^2 dx]^{1/2}$$

by Schwarz's inequality. Dividing each side by

$$[\int |f+g|^2 dx]^{1/2}$$

we get

$$[\int |f+g|^2 dx]^{1/2} \le [\int |f|^2 dx]^{1/2} + [\int |g|^2 dx]^{1/2}$$

and this completes the proof.

8.6 FOURIER SERIES

1. Integrable Functions:

Let f be a measurable function defined and integrable on a set $E \subset [a, b]$. Then we say that $f \varepsilon L_1(E)$. Similarly, we write $f \varepsilon L_2(E)$ when f is square integrable on E or f^2 is integrable on E. Thus $L_1(E)$ and $L_2(E)$ are spaces of integrable and square integrable functions respectively. If $E = [a, b]$ we write $f \varepsilon L_1 [a, b]$ and $f \varepsilon L_2 [a, b]$ respectively.

2. Periodic Functions:

Let $f : R \to R$ be a function such that

$$f(x + \alpha) = f(x), \quad \alpha > 0 \tag{62}$$

Then f is said to be a periodic function of period α. Obviously,

$$f(x + n\alpha) = f(x) \text{ for } n = 1, 2, 3, \ldots$$

Let us consider the periodic functions f when $\alpha = 2\pi$. All trigonometric functions fall in this category. Any periodic function f of period 2π is expressed as the trigonometric series

$$f(x) = \frac{1}{2} a_0 + \sum_{n=1}^{\infty} [a_n \cos nx + b_n \sin nx)] \tag{63}$$

where a_0, a_n and b_n are constants.

If $f \varepsilon L_1 [-\pi, \pi]$ the series defined by (63) is called *Fourier series* of f in the interval $[-\pi, \pi]$ and the respective coefficients are given by

$$a_n = \frac{1}{\pi} \int_{-\pi}^{\pi} f(x) \cos nx \, dx, \quad n = 0, 1, 2, 3, \ldots \tag{64a}$$

$$b_n = \frac{1}{\pi} \int_{-\pi}^{\pi} f(x) \sin nx \, dx, \quad n = 1, 2, 3, \ldots \tag{64b}$$

For any $f \varepsilon L_1 [-\pi, \pi]$ there is nothing to suggest that the series (63) is convergent. In fact several cases are possible:

(i) The series may not converge at all,

(ii) The series may converge but it may not converge to f,

(iii) The series converges to f.

However if $f \varepsilon L_2 [-\pi, \pi]$, then the Fourier series of f converges to f; but if $f \varepsilon L_1 [-\pi, \pi]$, or f is continuous then the arithmetic mean of the partial sums of the Fourier series of f satisfies some convergence criterion. We shall not prove these facts in this introductory book.

We have

8.6.1. Theorem

Let $f \varepsilon L_1 [-\pi, \pi]$ be periodic. Let its Fourier series be given by

$$f(x) = \frac{1}{2} a_0 + \sum_{n=1}^{\infty} (a_n \cos nx + b_n \sin nx) \qquad (65)$$

Then

$$a_n = \frac{1}{\pi} \int_{-\pi}^{\pi} f(x) \cos nx \, dx$$

$$b_n = \frac{1}{\pi} \int_{-\pi}^{\pi} f(x) \sin nx \, dx$$

Proof: We have that $f \varepsilon L_1 [-\pi, \pi]$, $\cos nx \varepsilon L_2 [-\pi, \pi]$, $\sin nx \varepsilon L_2 [-\pi, \pi]$ and the series on the right of (65) is uniformly convergent. Multiplying both the sides of (65) by $\cos nx$ and integrating with respect to x from $-\pi$ to π we get

$$\int_{-\pi}^{\pi} f(x) \cos nx \, dx = a_n \int_{-\pi}^{\pi} \cos^2 nx \, dx$$

as $$\int_{-\pi}^{\pi} \cos nx \sin kx \, dx = 0, \; k = 1, 2, 3, \ldots,$$

$$\int_{-\pi}^{\pi} \cos nx \cos kx \, dx = 0 \text{ for } k \neq n$$

Therefore

$$a_n = \frac{1}{\pi} \int_{-\pi}^{\pi} f(x) \cos nx \, dx, \; n = 0, 1, 2, 3, \ldots$$

Similarly, multiplying (65) by $\sin nx$ and integrating with respect to x from $-\pi$ to π we obtain the value of b_n for $n = 1, 2, 3, \ldots$

The uniform convergence of the series ensures the term by term integration in both the cases.

8.6.2. Corollary

If f is an even function, i.e. $f(-x) = f(x)$, $f \varepsilon L_1 [-\pi, \pi]$ and is periodic in this interval, then the Fourier series of f is

$$\frac{1}{2} a_0 + \sum_{r=1}^{\infty} a_n \cos nx \tag{66}$$

where

$$a_n = \frac{2}{\pi} \int_0^{\pi} f(x) \cos nx \, dx \tag{67}$$

Proof: Let F_e be an even function defined on R. Then we have

$$\int_{-a}^{a} F_e(x)\, dx = 2 \int_0^{a} F_e(x)\, dx$$

Again, if F_0 is an odd function, then

$$\int_{-a}^{a} F_0(x)\, dx = 0$$

It is also known that the product (i) of two odd functions is even, (ii) of one odd and one even function is odd; and (iii) of two even functions is even. As $\cos nx$ is even and $\sin nx$ is odd the integral on the right of (64a) is equivalent to

$$\frac{2}{\pi} \int_{-\pi}^{\pi} f(x) \cos nx \, dx$$

and on the right of (64b) is zero for an even function f. This yields the result.

8.6.3. Corollary

If f is an odd function, i.e. $f(-x) = -f(x)$, $f \varepsilon L_1 [-\pi, \pi]$, and is periodic in this interval, then the Fourier series of f is

$$\sum_{n=1}^{\infty} b_n \sin nx \tag{68}$$

where

$$b_n = \frac{2}{\pi} \int_0^{\pi} f(x) \sin nx \, dx. \tag{69}$$

The proof is similar to that given in the previous case.

8.6.4. Corollary

If f is a periodic function of period $2l$, l is finite, and $f = L_1[-l, l]$, then the Fourier series of f is given by

$$\frac{1}{2} a_0 + \sum_{n=1}^{\infty} \left[a_n \cos\left(\frac{n\pi x}{l}\right) + b_n \sin\left(\frac{n\pi x}{l}\right) \right] \tag{70}$$

where

$$a_n = \frac{1}{l} \int_{-l}^{l} f(x) \cos\left(\frac{m\pi x}{l}\right) dx, \; n = 0, 1, 2, 3, \ldots \tag{71a}$$

$$b_n = \frac{1}{l} \int_{-l}^{l} f(x) \sin\left(\frac{n\pi x}{l}\right) dx, \; n = 1, 2, 3, \ldots \tag{71b}$$

Proof: This corollary can be derived from the main theorem by replacing x by y and then putting

$$y = \frac{\pi x}{l}$$

8.6.5.

Let $\{g_1, g_2, g_3, \ldots\}$ be a set of real-valued functions, each defined on a closed interval $[a, b]$, with the property that $g_m g_n \in L_1$ for all m, n. This, of course, holds if each $g_n \in L_2$. Then such a set is said to be an *orthogonal set of functions* if

$$\int_a^b g_m(x)\, g_n(x)\, dx = 0 \; (m \neq n).$$

A set of functions $\{g_1, g_2, \ldots\}$ is called an *orthonormal set* if $g_m g_n \in L_1$ and

$$\int_a^b g_m(x)\, g_n(x)\, dx = \begin{cases} 0, & m \neq n \\ 1, & m = n. \end{cases}$$

8.6.6. Examples

(a) Observe that

$$\int_{-\pi}^{\pi} \sin mx \sin nx \, dx = \begin{cases} 0, & m \neq n \\ \pi, & m = n. \end{cases}$$

Therefore $\{\sin mx, m = 1, 2, \ldots\}$ is an orthogonal set and

$$\left\{\frac{\sin x}{\sqrt{\pi}}, \frac{\sin 2x}{\sqrt{\pi}}, \ldots.\right\}$$

is orthonormal.

(b) We have

$$\int_{-\pi}^{\pi} \cos mx \sin nx \, dx = 0, \; m \neq n.$$

Hence

$$\{1, \cos x, \sin x, \cos 2x, \sin 2x, \ldots\}$$

is orthogonal and the corresponding orthonormal set is

$$\left\{\frac{1}{\sqrt{2\pi}}, \frac{\cos x}{\sqrt{\pi}}, \frac{\sin x}{\sqrt{\pi}}, \frac{\cos 2x}{\sqrt{\pi}}, \frac{\sin 2x}{\sqrt{\pi}}, \ldots\right\}.$$

Recall that for $f \in L_1$, the Fourier series of f is given by (63) and its Fourier coefficients are given by (64a) and (64b).

In fact, if $\{g_1, g_2, \ldots\}$ is a set of orthogonal functions, each defined on $[a, b]$ and if f is a given function which can be represented in terms of g_i's by a convergent series of the form

$$f(x) = \sum_{n=1}^{\infty} c_n g_n(x) = c_1 g_1(x) + c_2 g_2(x) + \ldots,$$

then this series is called a *generalised* Fourier series of f.

We shall close this section with some examples.

Examples: 1. Obtain the Fourier series for $f(x) = e^{-x}$ in the interval $[0, 2\pi]$.

Solution: Let

$$e^{-x} = \frac{1}{2} a_0 + \sum_{r=1}^{\infty} (a_r \cos rx + b_r \sin rx)$$

Then

$$a_0 = \frac{1}{\pi} \int_0^{2\pi} e^{-x}\, dx = \frac{1}{\pi}(1 - e^{-2\pi})$$

$$\pi\, a_r = \int_0^{2\pi} e^{-x} \cos rx\, dx,\ \pi\, b_r = \int_0^{2\pi} e^{-x} \sin r x\, dx$$

So

$$\pi\, a_r = (1 - e^{-2\pi}) - \pi\, r\, b_r$$

Also

$$\pi\, b_r = \pi\, r\, a_r$$

Hence

$$a_r = \frac{(1 - e^{-2\pi})}{\pi\,(1 + r^2)},\ b_r = \frac{r}{\pi} \frac{(1 - e^{-2\pi})}{1 + r^2}$$

Thus, the Fourier series of e^{-x} is given by

$$e^{-x} = \frac{1}{\pi}(1 - e^{-2\pi})\left[\frac{1}{2} + \sum_{r=1}^{\infty} \frac{1}{r^2 + 1}(\cos r\pi + r \sin r\pi)\right]$$

2. Find the Fourier series of the periodic function $f(x)$ with period 2π defined by

$$f(x) = x,\ -\pi < x \leq \pi.$$

We have

$$a_0 = \frac{1}{\pi} \int_{-\pi}^{\pi} x dx = \frac{1}{\pi} \frac{x^2}{2} \Big|_{-\pi}^{\pi} = 0.$$

$$a_k = \frac{1}{\pi} \int_{-\pi}^{\pi} x \cos k x\, dx$$

$$= \frac{1}{\pi}\left[x \frac{\sin kx}{k} \Big|_{-\pi}^{\pi} - \frac{1}{k} \int_{-\pi}^{\pi} \sin kx dx \right] = 0$$

$$b_k = \frac{1}{\pi} \int_{-\pi}^{\pi} x \sin kx\, dx$$

$$= \frac{1}{\pi}\left[-x\frac{\cos kx}{k}\Big|_{-\pi}^{\pi} + \frac{1}{k}\int_{-\pi}^{\pi}\cos kx dx\right]$$

$$= (-1)^{k+1}\frac{2}{k}.$$

Hence the series is

$$f(x) = 2\left[\frac{\sin x}{1} - \frac{\sin 2x}{2} + \frac{\sin 3x}{3} - \ldots (-1)^{k+1}\frac{\sin kx}{k} + \ldots\right]$$

3. Find the Fourier series of a periodic function $f(x)$ of period 2π defined by

$$f(x) = \begin{cases} -x, & -\pi \le x \le 0 \\ x, & 0 < x \le \pi \end{cases}$$

$$a_0 = \frac{1}{\pi}\int_{-\pi}^{\pi} f(x)\, dx = \frac{1}{\pi}\left[\int_{-\pi}^{0}(-x)\, dx + \int_{0}^{\pi} x dx\right] = \pi.$$

$$a_k = \frac{1}{\pi}\left[\int_{-\pi}^{0}(-x)\cos kx dx + \int_{0}^{\pi} x\cos kx dx\right]$$

$$= \frac{1}{\pi}\left[-\frac{x\sin kx}{k}\Big|_{-\pi}^{0} + \frac{1}{k}\int_{-\pi}^{0}\sin kx dx + \frac{x\sin kx}{k}\Big|_{0}^{\pi} - \frac{1}{k}\int_{0}^{\pi}\sin kx dx\right]$$

$$= \frac{1}{\pi k}\left[-\frac{\cos kx}{k}\Big|_{-\pi}^{0} + \frac{\cos kx}{k}\Big|_{0}^{\pi}\right]$$

$$= \frac{2}{\pi k^2}(\cos k\pi - 1) = \begin{cases} 0, & k \text{ even} \\ -\dfrac{4}{\pi k^2}, & k \text{ odd} \end{cases}$$

$$b_k = \frac{1}{\pi}\left[\int_{-\pi}^{0}(-x)\sin kx dx + \int_{0}^{\pi} x\sin kx dx\right] = 0.$$

Thus the Fourier series is given by

$$f(x) = \frac{\pi}{2} - \frac{4}{\pi}\left[\frac{\cos x}{1^2} + \frac{\cos 3x}{3^2} + \ldots + \frac{\cos (2p+1)x}{(2p+1)^2} + \ldots\right]$$

4. Find the Fourier series of a periodic function $f(x)$ of period 2π defined by

$$f(x) = \begin{cases} -1, & -\pi < x < 0 \\ 1, & 0 \le x \le \pi \end{cases}$$

$$a_0 = \frac{1}{\pi}\int_{-\pi}^{\pi} f(x)\,dx = \frac{1}{\pi}\left[\int_{-\pi}^{0}(-1)\,dx + \int_{0}^{\pi} dx\right] = 0.$$

$$a_k = \frac{1}{\pi}\left[\int_{-\pi}^{0}(-1)\cos kx\,dx + \int_{0}^{\pi}\cos kx dx.\right]$$

$$= -1.\frac{\sin kx}{k\pi}\Big|_{-\pi}^{0} + \frac{\sin kx}{k\pi}\Big|_{0}^{\pi} = 0$$

$$b_k = \frac{1}{\pi}\left[\int_{-\pi}^{0}(-1)\sin kxdx + \int_{0}^{\pi}\sin kxdx\right]$$

$$= \frac{1}{\pi}\left[\frac{\cos kx}{k}\right]\Big|_{-\pi}^{0} - \frac{\cos kx}{k}\Big|_{0}^{\pi}$$

$$= \frac{2}{\pi k}(1 - \cos \pi k) = \begin{cases} 0, & k \text{ even} \\ \dfrac{4}{\pi k}, & k \text{ odd} \end{cases}$$

Hence the series is

$$f(x) = \frac{4}{\pi}\left[\frac{\sin x}{1} + \frac{\sin 3x}{3} + \ldots + \frac{\sin (2p+1)x}{2p+1} + \ldots\right]$$

5. Find the Fourier series of a periodic function $f(x)$ of period 2π defined by

$$f(x) = x^2, \; -\pi \le x \le \pi.$$

$$a_0 = \frac{1}{\pi}\int_{-\pi}^{\pi} x^2 dx = \frac{1}{\pi}\frac{x^3}{3}\Big]_{-\pi}^{\pi} = \frac{2\pi^2}{3}$$

$$a_k = \frac{1}{\pi}\int_{-\pi}^{\pi} x^2 \cos kx dx$$

$$= \frac{1}{\pi}\left[\frac{x^2 \sin kx}{k}\Big|_{-\pi}^{\pi} - \frac{2}{k}\int_{-\pi}^{\pi} x \sin kx dx\right]$$

$$= -\frac{2}{\pi k}\left[-\frac{x \cos kx}{k}\Big|_{-\pi}^{\pi} + \frac{1}{k}\int_{-\pi}^{\pi} \cos kx dx\right]$$

$$= \frac{4}{\pi k^2}(\pi \cos k\pi) = \begin{cases} \frac{4}{k^2}, & k \text{ even} \\ -\frac{4}{k^2}, & k \text{ odd} \end{cases}$$

$$b_k = \frac{1}{\pi}\int_{-\pi}^{\pi} x^2 \sin kx dx$$

$$= \frac{1}{\pi}\left[\frac{-x^2 \cos kx}{k}\Big|_{-\pi}^{\pi} + \frac{2}{k}\int_{-\pi}^{\pi} x \cos kx dx\right]$$

$$= \frac{2}{\pi k}\left[\frac{x \sin kx}{k}\Big|_{-\pi}^{\pi} - \frac{1}{k}\int_{-\pi}^{\pi} \sin kx dx\right] = 0.$$

Hence

$$x^2 = \frac{\pi^2}{3} - 4\left(\frac{\cos x}{1} - \frac{\cos 2x}{2^2} + \frac{\cos 3x}{3^2} - \dots\right)$$

6. Find the Fourier series of a periodic function $f(x)$ of period 2π defined by

$$f(x) = \begin{cases} 0, & -\pi < x \le 0 \\ x, & 0 < x \le \pi \end{cases}$$

$$a_0 = \frac{1}{\pi}\int_{-\pi}^{\pi} f(x) dx = \frac{1}{\pi}\left[\int_{0}^{\pi} 0.dx + \int_{0}^{\pi} x dx\right]$$

$$= \frac{1}{\pi}\frac{\pi^2}{2} = \frac{\pi}{2}.$$

$$a_k = \frac{1}{\pi}\int_0^{\pi} x\cos kx dx = \frac{1}{\pi}\left[\frac{x\sin kx}{k}\Big|_0^{\pi} - \frac{1}{k}\int_0^{\pi}\sin kx dx\right]$$

$$= \frac{1}{\pi k}\frac{\cos kx}{k}\Big|_0^{\pi} = \begin{cases} -\dfrac{2}{\pi k^2}, & k \text{ odd} \\ 0, & k \text{ even} \end{cases}$$

$$b_k = \frac{1}{\pi}\int_0^{\pi} x\sin kx dx$$

$$= \frac{1}{\pi}\left[-\frac{x\cos kx}{k}\Big|_0^{\pi} + \frac{1}{k}\int_0^{\pi}\cos kx dx\right]$$

$$= -\frac{\pi}{\pi k}\cos k\pi = \begin{cases} \dfrac{1}{k}, & k \text{ odd} \\ -\dfrac{1}{k}, & k \text{ even} \end{cases}$$

Hence

$$f(x) = \frac{\pi}{4} - \frac{2}{\pi}\left(\frac{\cos x}{1^2} + \frac{\cos 3x}{3^2} + \ldots\right) + \left(\frac{\sin x}{1} - \frac{\sin 2x}{2} + \frac{\sin 3x}{3} - \ldots\right).$$

EXERCISES

1. For $E \subset R$ let $-E = \{x \mid -x \,\varepsilon\, E\}$ show that $m_0(E) = m_0(-E)$.
2. Let $E \subset M$ where M is measurable and $m(M) < \infty$. Show that E is measurable if and only if $m(M) = m_0(E) + m_0(M - E)$.
3. Let E be a set, x_0 a fixed real number, and U a set such that $U = \{x + x_0 \mid x \,\varepsilon\, E\}$
   Then show that
   $$m_0(E) = m_0(U).$$
4. Prove that the set of all irrational numbers in [0. 1] is measurable, find its measure.

5. Let E be the set of all numbers which can be represented in the form

$$\frac{a_1}{5} + \frac{a_2}{5^2} + \ldots. + \frac{a_n}{5^n} + \ldots$$

where $a_n = 0$ or 4 for each n. Then prove that

$$m(E) = 0.$$

6. Let E be defined as

$$E = \bigcup_{n=1}^{\infty} \left[n - \frac{1}{3^n}, n \right]$$

then show that E is measurable (although unbounded) and $m(E) = 1/2$.

7. Prove that the function f where $f(0) = 5$ and $f(1) = 7$;

$$f(x) = \frac{1}{x} \text{ for } 0 < x < 1 \text{ is measurable.}$$

8. Show that monotone functions are measurable.

9. Let f be defined on $[0, 1]$ by $f(0) = 0$. $f(x) = x \sin 1/x$ for $x > 0$. Then find the measure of the set

$$\{x : f(x) \geq 0\}$$

10. If f_1 and f_2 are measurable functions on $[a, b]$ then show that the functions max (f_1, f_2) and min (f_1, f_2) are also measurable.

11. Show that the function f, where

$$f(x) = \begin{cases} x + 5 & \text{for } x < -1, \\ 2 & \text{for } -1 < x < 0, \\ x^2 & \text{for } x \geq 0 \end{cases}$$

is measurable.

12. Prove that the step function $[x]$ is measurable, where [ ] denotes the largest integer contained in x.

13. Let f be finite-valued and measurable on $[a, b]$. Then prove that $|f|^\alpha$ for $\alpha \geq 0$ is measurable.

14. Find $\int_0^1 f dx$ where $f(x) = x^{-1/3}$ for $0 < x \leq 1$ and $f(0) = 0$

15. Find $\int_a^b f\,dx$ where $f(x) = 1$ for x is rational and $f(x) = 2$ for x is irrational.

16. Find $\int_{-1}^{1} f\,dx$ where $f(x) = (1 + x^2)^{-1}$ for $x \,\varepsilon\, [-1, 1]$.

17. Show that $\int_1^{\infty} f$ is unbounded for $f(x) = \frac{1}{x}$.

18. Expand the following functions in Fourier series:

(i) $f(x) = x^n, -\pi < x < \pi$, n is a positive integer.

(ii) $f(x) = |\,x\,|, -\pi < x < \pi$.

(iii) $f(x) = \begin{cases} 1 + \dfrac{2x}{\pi}, & -\pi < x \le 0 \\ 1 - \dfrac{2x}{\pi} & 0 \le x < \pi \end{cases}$

19. $f(x) = 9 - x^2, -3 < x < 3$

20. $f(x) = \cos \alpha x, \ -\pi < x < \pi, \ 0 < \alpha < 1$.

BIBLIOGRAPHY

W. A Albrechet and A. H. Smith, *Fundamental Concepts of Analysis*, Prentice Hall of India Pvt. Ltd. New Delhi (1969).

B. Choudhary, *The Elements of Complex Analysis*, Wiley Eastern Ltd. New Delhi (1983).

B. Choudhary and S. Nanda, *Functional Analysis with Applications*, Wiley Eastern Ltd. New Delhi (1989).

E. Hewitt and K. Stromberg, *Real and Abstract Analysis*, Springer Publishing Co. Inc. New York (1965).

I. N. Herstein, *Topics in Algebra*, Blaisdell Publishing Co. New York (1964).

S. C. Malik, *Mathematical Analysis*, Wiley Eastern Ltd. New Delhi (1982).

S. Nanda, *Number Theory*, Allied Publishers, New Delhi (1985).

S. Nanda, Topics in Algebra, Tata Mc-Graw Hill Publishing Co. New Delhi (1989).

Sribatsa Nanda and Sudarsan Nanda, *General Topology*, Macmillan India (1990).

G. S. Pandey and V. P. Saxena, *Calculus of Two and More Variables*, Wiley Eastern Ltd. New Delhi (1987).

G. S. Pandey, V. P. Saxena and R. L. Mahajan, *Calculus of One Variable*, Wiley Eastern Ltd. New Delhi (1988).

W. Rudin, *Real and Complex Analysis*, Mc-Graw Hill Book Co. New York (1974).

W. Rudin, *Principles of Mathematical Analysis*, Mc-Graw Hill Kogakusha, Ltd. (1976).

H. L. Royden, *Real Analysis*, The Macmillan Co. New York (1963).

G. F. Simmons, *Introduction to Topology and Modern Analysis*, Mc-Graw-Hill Book Co. New York (1974).

Shantinarayan, *A Course of Mathematical Analysis*, S. Chand and Co. Ltd. New Delhi (1979).

S. C. Saxena and S. M. Shah, *Introduction to Real Variable Theory*, Prentice Hall (India), New Delhi (1987).

Subject Index

O

P

R

S

T

U

W

9 788177 640625

Printed at Repro India Ltd.